Machine Learning in Geomechanics 1

SCIENCES

Mechanics, Field Director – Gilles Pijaudier-Cabot

Geomechanics, Subject Head – Gioacchino Viggiani

Machine Learning in Geomechanics 1

Overview of Machine Learning, Unsupervised Learning, Regression, Classification and Artificial Neural Networks

Coordinated by
Ioannis Stefanou
Félix Darve

ISTE

WILEY

First published 2024 in Great Britain and the United States by ISTE Ltd and John Wiley & Sons, Inc.

ISTE Ltd
27-37 St George's Road
London SW19 4EU
UK

www.iste.co.uk

John Wiley & Sons, Inc.
111 River Street
Hoboken, NJ 07030
USA

www.wiley.com

Library of Congress Control Number: 2024943344

British Library Cataloguing-in-Publication Data
A CIP record for this book is available from the British Library
ISBN 978-1-78945-192-4

ERC code:
PE6 Computer Science and Informatics
PE6_7 Artificial intelligence, intelligent systems, natural language processing
PE10 Earth System Science
PE10_19 Planetary geology and geophysics

Contents

Preface

Ioannis STEFANOU[1] and Félix DARVE[2]

[1] *GeM, UMR 6183, CNRS, Ecole Centrale Nantes, Nantes Université, France*

[2] *3SR Laboratory, CNRS, Grenoble INP – ENSE3, Université Grenoble Alpes, France*

When discussing artificial intelligence (AI), some basic questions immediately emerge: What is AI? How does it work?

Behind this well-known term, a collection of methods of applied mathematics allow the computer to learn and identify patterns in data. This collection of methods is called machine learning (ML) and it is the target of these two volumes, which were authored for the 2023 ALERT Geomaterials Doctoral School.

In combination with the tremendous increase in the computational power, ML has led to incredible achievements in many disciplines of science and technology. These achievements were that striking that some researchers believe that ML could become a turning point for humanity, as the discovery of fire was for our far ancestors!

Until the 1960s, the scientific development has been characterized by the so-called "linear physics" and by modeling represented by analytical equations solved explicitly by the available mathematical tools giving rise to analytical solutions. The field of problems that can be solved in this way is, of course, precious, but very limited.

Machine Learning in Geomechanics 1,
coordinated by Ioannis STEFANOU and Félix DARVE.

Then the numerical revolution, based on powerful numerical methods and computers, allowed to solve numerically a large variety of problems, which can be described by a known system of equations. Many limits of this methodology are known due to the abundance of nonlinear processes in nature, chaos and complexity. In any case, numerical analysis, another branch of applied mathematics, has immensely enlarged the class of problems that can be solved today.

However, the numerical solutions of these sets of nonlinear equations can be computationally very intensive or even impossible. Moreover, many problems in engineering are difficult to describe by a set of equations. ML tools provide promising methods for addressing both those problems.

Another aspect of ML algorithms is their ability to solve very complex problems in a "creative" manner. One characteristic example of creativity was shown when the machine beat the world champion of Go, which was invented a long time ago in China. Differently from chess, in which the computer can predict the game evolution several moves in advance, in Go the number of possible moves is extremely large (higher than the number of atoms in the known universe). Therefore, it is necessary to follow creative strategies. Indeed, the machine, "AlphaGo", has shown that it is able to carry out novel strategies that surprised even the best human players in the world.

All these methods of ML give new powerful tools to scientists and engineers and open new perspectives in geomechanics. The target of these two volumes is to demystify ML to present its main methods and to show some examples of applications in mechanics and geomechanics. Most of the chapters of the volume were edited to provide a pedagogical introduction to the most important methods of ML and to uncover the fundamental notions behind them.

Volume I contains the following five chapters:

The first chapter, "Overview of machine learning in geomechanics", is the introductory chapter of this volume. In this chapter, we explain how the machine can learn, show a classification of the main methods in ML, outline some applications of ML in geomechanics and highlight its limitations.

The second chapter, "Introduction to regression methods", focuses on regression, which is one of the fundamental pillars of supervised ML. In this chapter, we introduce the essential concepts in regression analysis and methods by providing hands-on, practical examples.

The target of the third chapter, "Unsupervised learning: Basic concepts and application to particle dynamics", is twofold. The first part of this chapter is devoted to the description of the basic concepts of the most popular techniques of unsupervised learning. The second part illustrates an application of unsupervised learning to the discovery of patterns in particles dynamics.

The fourth chapter, "Classification techniques in machine learning", aims to describe what the problem of classification in ML is and illustrates some of the methods used for solving it, without resorting to artificial neural networks (ANNs). Hands-on examples are given and active learning is discussed.

The fifth chapter, "Artificial neural networks: Learning the optimum statistical model from data", provides a comprehensive introduction to ANNs. Several hands-on examples are given to help the reader grasp the main ideas.

Volume II is organized as follows:

The first chapter, "Data-driven modeling in geomechanics", presents the theoretical framework of the so-called data-driven computational mechanics. Furthermore, it shows some of its applications for the solution of problems involving Cauchy and Cosserat continua with elastic and inelastic materials, which, naturally, represent common descriptions of geomaterials.

The second chapter, "Bayesian inference in geomechanics", is intended to provide a concise exploration of Bayesian inference and demonstrates how recent advancements in ML can assist in efficient Bayesian inference within the realm of geomechanics applications.

The third chapter, "Physics-informed and thermodynamics-based neural networks", shows how to inject prior knowledge into deep learning algorithms. Using various examples, we present physics-informed neural networks for the discovery of partial differential equations and thermodynamics-based ANNs for the discovery of constitutive models of complex, inelastic materials.

The fourth chapter, "Introduction to reinforcement learning with applications in geomechanics", presents the basic concepts of reinforcement learning, which enables the development of software agents that are capable of making optimal decisions in dynamic and uncertain environments. The chapter closes with two applications of reinforcement learning in geomechanics.

The fifth chapter, "Artificial neural networks: basic architectures and training strategies", presents more architectures of ANNs and discusses training strategies.

We deeply thank all the authors of the volumes for their comprehensive contributions and their effort to present complex notions in a pedagogical manner. We also deeply thank ALERT Geomaterials for the organization of this doctoral school and all students for their active participation. We hope that the chapters provide a valuable introduction to machine learning in geomechanics.

Ioannis STEFANOU
Félix DARVE

June 2024

1

Overview of Machine Learning in Geomechanics

Ioannis Stefanou
GeM, UMR 6183, CNRS, Ecole Centrale Nantes,
Nantes Université, France

1.1. What exactly is machine learning?

One of the best methods for making a puppy learn its name is to play the name game:

1) Take your puppy to a quiet place and gain its attention by calling its name in a happy and reassuring tone (e.g. “LoupI look at me!” or “LoupI come here!”).

2) As soon as you have gained its attention, reward it with a treat to encourage its good behavior.

3) Repeat this process several times until your puppy has learned to respond to its name.

For a color version of all the figures in this chapter, see www.iste.co.uk/stefanou/machine1.zip.

Machine Learning in Geomechanics 1,
coordinated by Ioannis Stefanou and Félix Darve.

In the above iterative process, the puppy learns from experience that reacting when you call its name is a good thing. By repeating this process, the puppy recognizes the pattern between similar, but not identical, vocal signals of its name (training data) and associates them with a specific behavior (task).

Many cognitive processes of humans involve a similar way for associating patterns with specific reactions; for instance, see how we learn to drive a car, speak a foreign language or even understand how a geomaterial behaves!

Machine learning (ML) is an anthropomorphic extension of the above process of "learning" to a computer program, that is, to the machine. By learning, the machine will progressively improve its performance in executing a specific task. According to Mitchell (1997), ML can be defined as follows:

> *A computer program is said to learn from experience E with respect to some class of tasks T and performance measure P, if its performance at tasks in T, as measured by P, improves with experience E.*

The whole set of examples that the computer will use during learning is the *training set*, while each training example is called a training instance or sample. The whole process of learning is called *training*. The training set is usually distinguished into *training data*, *validation data* and *test data*. More details about those sets and their importance will be given in section 1.5 and in Chapters 2 and 5 of volume 1.

A popular example showing the power of ML was *AlphaGo*, a ML program, which beat the human world champion of Go (see also (Silver et al. 2016) and Chapter 4 of volume 2 on *reinforcement learning*). Go is a strategy board game for two players in which the aim is to cover more territory than the opponent (see Figure 1.1). Despite its relatively simple rules, Go is extremely complex. The *state-space complexity* of Go, that is, the number of legal game positions reachable from the initial position of the game (Allis 1994), is of the order of 10^{170} (Tromp and Farnebäck 2007), which is much higher than that of chess, which is of the order of 10^{44} (Shannon 1950), and certainly higher than the number of atoms in the universe. This makes Go a great challenge for ML algorithms.

Figure 1.1. *Eighteen times Go world champion Lee Sedoll playing with the machine (courtesy: Lee Jin-man/AP, The New Yorker, https://www.newyorker.com/tech/annals-of-technology/alphago-lee-sedol-and-the-reassuring-future-of-humans-and-machines)*

Applying the above-mentioned definition in an ML program for Go, the performance (P) will be measured by the ability of the machine to beat its opponent by gaining experience (E) by executing multiple times the task (T), which is playing the game with itself. Experience can be also obtained by processing existing games of other (human) players.

Another recent popular example of ML is the *ChatGPT* chatbot developed by OpenAI. ChatGPT was released in November 2022 and provoked a massive public reaction. Let us see what it answers when we ask it to describe itself and what it knows about ALERT Geomaterials.

Ioannis: *Describe yourself.*

ChatGPT: *I am ChatGPT, an AI language model developed by OpenAI. I have been trained on a vast amount of text data from the internet, books and other sources. My purpose is to assist* *users in generating human-like text responses and providing information on a wide range of topics.*

I don't have a physical form or personal experiences like humans do, but I can understand and generate text in multiple languages. I strive to provide helpful and accurate responses to the best of my abilities, drawing upon the knowledge I have acquired during my training.

I can engage in conversations, answer questions, offer explanations and assist with various tasks. However, please keep in mind that while I strive for accuracy, I may occasionally generate incorrect or incomplete information. It's always a good idea to verify important details from reliable sources.

Overall, I am here to assist and provide information to the best of my abilities. Feel free to ask me anything you'd like to know or engage in a conversation on any topic of your interest.

Ioannis: *How do you work?*

ChatGPT: *As an AI language model, my functioning can be divided into two main phases: training and inference.*

Training: During the training phase, large amounts of text data are used to train the model. In the case of GPT models like ChatGPT, this training involves a process called unsupervised learning. The model learns to predict what comes next in a given sequence of text, based on the patterns and relationships it discovers in the training data. By doing this over a vast dataset, the model gradually learns grammar, facts, reasoning abilities and the nuances of human language.

Inference: Once the training is complete, the model moves into the inference phase, where it can generate responses and interact with users. Here's a simplified overview of how it works during inference:

Input: When you provide a prompt or a message, the model receives it as input. The input can be a question, a statement or any text you want to generate a response to.

Tokenization: The model breaks down the input text into smaller units called tokens. These tokens can represent words, subwords or characters. Tokenization helps the model understand the structure and meaning of the input.

Processing: The tokenized input is then passed through several layers of neural networks. These layers are designed to understand the relationships between the tokens and capture the context and meaning of the text.

Generation: Once the input is processed, the model generates a response. It predicts the most likely next tokens based on the context and the patterns it has learned during training. The generated tokens form the model's response.

Iteration: The process of tokenization, processing and generation can happen in an iterative manner. The generated response can be further used as input, creating a back-and-forth conversation or a continuous interaction.

It's important to note that while ChatGPT can generate impressive responses, it may sometimes produce incorrect or nonsensical answers. It doesn't have real-world experiences or access to current information beyond its knowledge cutoff date. It's always a good idea to fact-check and verify important information from reliable sources.

Ioannis: *What is ALERT Geomaterials in summary?*

ChatGPT: *In summary, ALERT Geomaterials is an international scientific association dedicated to the study of the mechanics and behavior of geomaterials, including soil, rock and other geological materials. It brings together researchers and professionals in the field of geomechanics and geotechnical engineering to promote collaboration, knowledge exchange* *and scientific advancements. ALERT Geomaterials organizes workshops, conferences and training courses, providing a platform for researchers to share their findings and discuss topics related to soil and rock mechanics, slope stability, geotechnical engineering, experimental testing and numerical modeling. The association also publishes a journal focused on energy geotechnics and geoenvironmental engineering. Overall, ALERT Geomaterials plays a crucial role in fostering research and innovation in the field of geomaterials.*

Impressive! At this point, it becomes quite tempting to let ChatGPT finish this chapter or even both volumes! However, I assure you, we took the necessary time to entirely draft these volumes word by word. Read carefully ChatGPT's answers. Are all of them correct?

It is straightforward to see how the above definition for ML applies to ChatGPT and to many other applications of ML, such as speech, handwriting, face and object recognition, voice-to-text and vice versa, translation, text auto-correction and auto-completion, spam filtering, computer games, self-driving cars, medicine, forecasting, banking, security, marketing, control problems, engineering and, of course, (geo-)mechanics.

ML is an evolving field of knowledge and involves a plethora of methods and combinations of those. In the next section, we will try to categorize and classify them in groups.

Having described and defined what ML is, it is natural to ask what is *artificial intelligence* (AI)? ML and AI are closely related and the latter is considered to include the former. The exact definition of AI seems to be a bit foggy for the time being and depends on how we define the terms "artificial" and "intelligence". In the author's opinion, it is easier to describe the characteristics of AI, rather than give a unique and exact definition of the term. Another example of a hard definition, in a totally different domain, is what is justice? It seems easier and more important to describe the characteristics of justice (e.g. equality for all) rather than give a precise definition of the term.

1.2. Classification of ML methods

There are numerous ML methods in the literature. Therefore, it is useful to classify them into different categories. Here, we follow the classification of (Géron 2019), who categorizes ML to *supervised learning* versus *unsupervised learning*, *batch learning* versus *online learning* and *instance-based learning* versus *model-based learning* methods. Of course, this is a rough classification and one method can combine different categories, as shown in Figure 1.2.

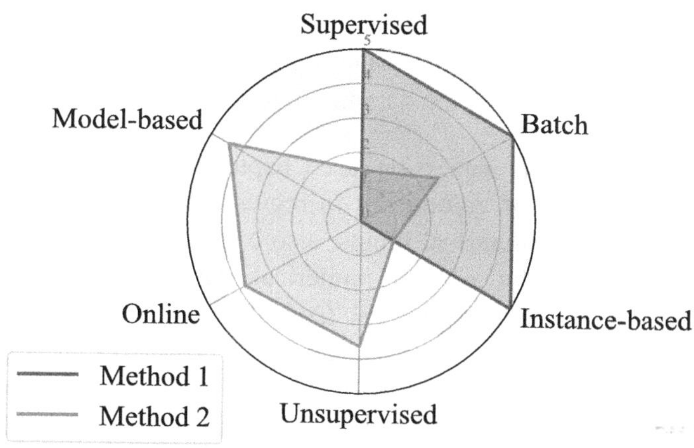

Figure 1.2. *Classification of ML methods. Method 1 is a supervised, batch, instance-based method, while Method 2 has components belonging to different categories*

1.2.1. *Supervised versus unsupervised ML*

In *supervised learning*, the training set includes the desired solutions/predictions, which are called *labels*.

For instance, imagine that your training set is hundreds of images with each one containing either a dog or a cat and that the training task is to distinguish the photos of dogs from those of cats. This is called a *classification* problem (see Chapter 4). If we give to the computer the information of which photos show dogs and which cats, or in other words, if we label the training set, then the learning is called supervised. If the training is successful, then the computer will have learned to identify cats from dogs even in a new set of photos that it has not processed during its training.

On the contrary, if the training set is unlabeled, then the learning is called *unsupervised learning*. In this case, the computer will eventually understand the pattern of the two different animals shown in the photos and it will be able to distinguish dogs from cats in an unsupervised manner (see also *clustering* in Chapter 3 of volume 1). Of course, the machine will not have learned to call a cat, cat and a dog, dog, because we have not given this additional information, but it will have identified their differences and separated the data into two different classes.

Another ML problem is *regression* (see Chapters 2 and 5 of volume 1 and Chapters 3 and 5 of volume 2). In regression, the training set contains one or several numerical inputs, also called *features*, and the task is to predict one or several numerical outputs, also called *predictors*, that depend on the inputs. Consider as an example the prediction of the stress response of a geomaterial, which as we know depends on several input parameters, such as the applied strain, available information about the evolution of its microstructure (e.g. the position and the velocity of the grains of a sand obtained by a discrete element method [DEM] analysis), history and/or other features. As the training set contains both the output and the input, the training for predicting the stresses based on the above-mentioned features is supervised. However, the identification/extraction by the machine of a representation of the most important features of the microstructure that are related to the prediction of the stress response is unsupervised. Examples of unsupervised methods in ML are *feature extraction*, *anomaly detection* and *dimensionality reduction*, in which the aim is to reduce the size of the training set without losing important information with respect to a specific task or measure and *data compression* (among others).

In Table 1.1, we provide a list of important supervised and unsupervised methods in ML. However, not all ML methods can be categorized to supervised and unsupervised. A notable example is *reinforcement learning* (RL), which does not require labeled data or a training set. For more details on RL, we refer to Chapter 4 of volume 2. Finally, when labels are not available for all the samples in the training set, we refer to *semisupervised learning*. Most semisupervised methods are a combination of supervised and unsupervised algorithms. An example of semisupervised learning is *active learning* (see Chapter 4 of volume 1 for more details).

ML method	Supervised	Unsupervised	Online	Presented in these volumes
Linear regression	✓			✓
Logistic regression	✓			✓
Polynomial regression	✓			✓
Lasso, Ridge	✓			✓
k-Nearest neighbors	✓			✓
Support vector machines (SVM)	✓			✓
Decision trees	✓			
Random forests	✓			
Artificial neural networks (ANNs)	✓		✓	✓
Autoencoders		✓	✓	✓
Clustering		✓		✓
Principal component analysis (PCA)		✓	✓*	
Locally linear embedding (LLE)		✓		
Reinforcement learning (RL)			✓	✓

Table 1.1. *Classification of some important methods in ML. The asterisk denotes that not all variations of the method fall into this category*

1.2.2. *Batch versus online ML*

Another manner to classify ML algorithms is based on whether they can improve their predictions by providing them with new data that may become available after the first training.

In *batch learning*, the machine has to be trained over a fixed training data set, without being able to add more data to the training set. Therefore, batch ML methods cannot improve their performance in a specific task by providing them with more data after the end of the training.

On the contrary, ML methods that support *online learning* allow to modify and increase the initial training set. Consequently, they are more flexible and suitable for large training sets, as they can be partially loaded into the memory of the computer and used whenever needed. Old data, over which the computer has been already trained, can also be erased to save space. Online learning is also very convenient when a constant flow of information exists, contrary to batch learning, in which the machine must be retrained over the whole data set. In other words, in batch learning, when new data become available the training set has to be updated and the training has to be repeated from scratch. This can be fine for some applications, but in many others it could have a very high computational and data storage cost.

Not all ML algorithms support online learning (see Table 1.1).

1.2.3. *Instance-based versus model-based ML*

One more way to categorize ML methods is based on the way data is learned in order to make predictions for data outside the training set. Two ML categories can be distinguished: *instance-based* (also known as *memory-based* or *lazy*) learning and *model-based* learning (also known as *physics-based* in some applications).

In instance-based learning, training data are simply interpolated in a high (usually) dimensional space. Then, new predictions are made based on how close or how similar new data are to those used for training. Similarity is measured on the basis of a distance measure, depending on the data and the problem at hand. In this sense, data are "memorized" by the machine, thus the term memory based. The quality of the predictions of data outside the training set is determined by how well new data are represented by the data of the training set. Instance-based approaches are straightforward to apply in any data set and they do not require any particular knowledge about the structure of the data or other characteristics that they might have. The hope of the user of instance-based methods is the machine will eventually identify by itself the hidden patterns in the data and give correct predictions even for data outside the training set. Examples of instance-based methods are *artificial neural networks* (ANNs), *decision trees*, *random forests*, k-*nearest neighbors* and many *clustering* techniques, *locally linear embedding* (LLE) and *principal component analysis* (PCA).

Despite the versatility and the many advantages of instance-based ML methods, they have an important drawback. In physics and engineering, we know that data have to respect at least some fundamental principles, for instance, the conservation of mass, energy and linear and angular momentum. Therefore, instance-/memory-based predictions that do not respect these conservation laws are unacceptable and can be even dangerous for applications. Data in this case have to be processed and learned by the machine under a model that includes the laws of physics.

Model-based ML adopts a model with some model parameters, which are optimized during training in order to optimally represent the data of the training set. Then the model is used to make predictions for unseen input data. Of course, if the adopted model is poor, the predictions will be poor as well.

Both instance-based and model-based approaches can introduce bias: the former because of limited data over which they were trained and the latter because of the model chosen. To fix the ideas, an example of instance-based and model-based ML is given in Figure 1.3. The training set contains the shear stress at failure (*predictor*) for a given normal stress (*feature*) of a series of experimental tests of a frictional interface. An instance-based method could give very poor predictions for unseen data, while a model-based prediction will be as good as the model is for describing the data. In this simple example, a Coulomb model was adopted for the model-based approach.

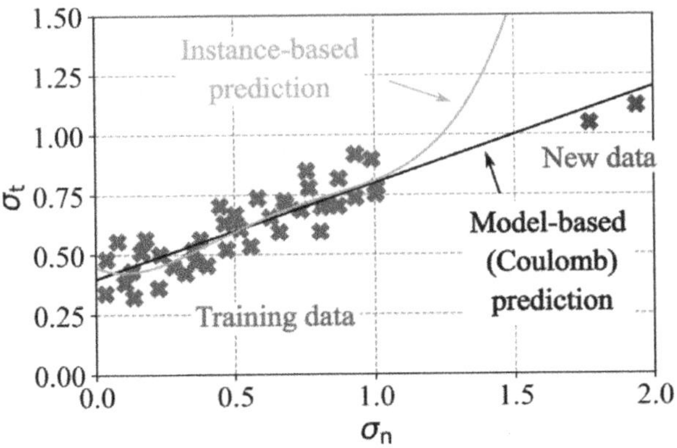

Figure 1.3. *Instance-based versus model-based ML for predicting failure (slip) of a frictional interface based on experimental data (fictitious)*

1.3. ML and geomechanics

The applications of ML in science are nowadays numerous and increasing. The same holds for applications of ML in geomechanics. It is out of the scope of this chapter to provide a comprehensive literature review, but it is worth mentioning some research directions in *constitutive modeling*, *geotechnics*, *geophysics* and *image correlation* that can be a starting point for the interested reader.

Geomaterials are among the most complex materials to study and model. The main reason is that their mechanical response is governed by multiphysics couplings at multiple spatial and temporal scales, which result in a macroscopic strongly nonlinear and dissipative response.

Empirical constitutive models are often used to describe the mechanical behavior of geomaterials. Classification and regression methods can be used for choosing the appropriate constitutive model that best fits the experimental data (see (Marques et al. 2022; Morand and Helm 2019; Guo et al. 2021; Schulte et al. 2023; Pouragha et al. 2020; Zhang et al. 2021a) and references therein). Another challenging task is to capture the macroscopic behavior of geomaterials based on the behavior of their microstructure and its evolution due to loading. Multiscale approaches can be employed for this purpose, requiring the solution of a boundary value problem (BVP) of an elementary cell of the microstructure at each point and time increment of the macroscopic analysis. Then, based on homogenization, the intrinsic constitutive behavior of the microstructure can be upscaled to the macroscopic level. For this purpose, mixed numerical schemes, such as the FE^2 method (Feyel 2003; Lloberas Valls et al. 2019; Eijnden et al. 2017) and FEM×DEM (Nitka et al. 2011; Nguyen et al. 2014) are often employed, depending on the nature of the microstructure. However, the computational cost of these methods is extremely high for real-scale applications, if not prohibitive. ML and ANN are one of the most promising ways to speed-up this multiscale process.

Recent works have shown that ANN can successfully encapsulate several aspects of the constitutive behavior of the underlying microstructure and provide the necessary information to the macroscopic scale with reduced calculation cost (see (Ghaboussi et al. 1991; Lefik and Schrefler 2003; Mozaffar et al. 2019; Liu and Wu 2019; Huang et al. 2020; Vlassis and Sun 2021; Zhou et al. 2021; Wang et al. 2019; Rocha et al. 2021; Becker et al. 2022; Wu et al. 2023; Sun et al. 2022; Zhang et al. 2023; Peng et al. 2021) to mention few). Going a step further, ANN can be designed in such a way to respect, by construction, the laws of physics (Karniadakis et al. 2021; Raissi et al. 2019), symmetries (Heider et al. 2020) and thermodynamics (Masi et al. 2021; Masi and Stefanou 2022, 2023; Hernandez et al. 2021) (see also Chapter 3 of volume 2). Once trained, these approaches can tremendously speed up the solution of difficult multiscale problems, guarantee the respect of the thermodynamic restrictions in their predictions and enable the extraction of the hidden state variables of the material. The latter can, in turn, shed light on the importance and the role of specific micromechanisms to the overall macroscopic behavior of complex (geo-)materials. The above collection of ML approaches is enriched by the so-called "data-driven" methods, which present an alternative formulation, whereby optimal material states are sought within a dataset that most closely satisfies momentum and energy conservation principles (Karapiperis et al. 2020; Kirchdoerfer and Ortiz 2016, 2018; Karapiperis et al. 2021) (see also Chapter 1 of volume 2).

Moving to applications of ML in geotechnics, according to the recent review of (Baghbani et al. 2022) (see also (Zhang et al. 2021b)), more than 1,200 articles can be found in the literature starting from the early 1990s. According to the same source, a net burst of production of scientific articles is observed after 2017. Note that more than half of these works use ANN. Several areas of geotechnical engineering are covered (see Figure 1.4), such as frozen soils and soil thermal properties, rock mechanics, subgrade soils and pavements, landslides (see (Tehrani et al. 2022) for a recent review), liquefaction, slope stability, shallow foundations, piles, tunneling and tunnel boring machines, dams and unsaturated soils, among others. In 2018, the increasing interest of the geotechnical community in ML led (Zhang and Liu 2022) to the creation of a new technical committee (TC) in the International Society for Soil Mechanics and Geotechnical Engineering (ISSMGE), entitled TC309 "Machine Learning and Big Data" (https://www.issmge.org/committees/technical-committees/impact-on-society/machine-learning, see also TC304).

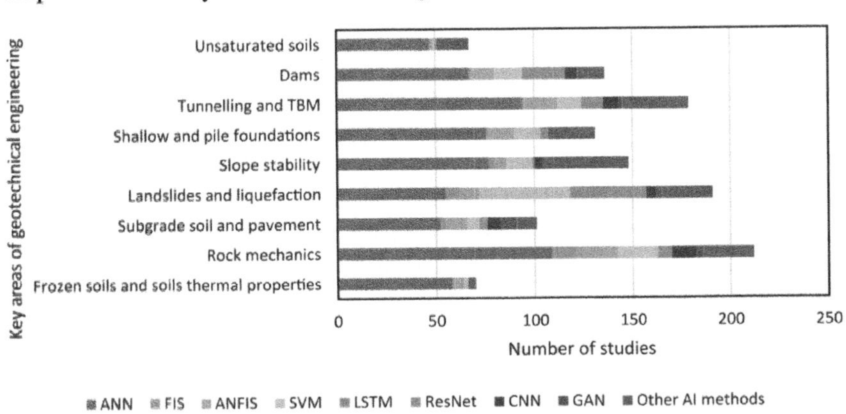

Figure 1.4. *Number of published articles using ML for geotechnical applications (slightly modified from Baghbani et al. (2022))*

ML also has promising applications in geophysics. For instance, it finds applications in geophysical exploration, reservoir engineering and drilling (Sircar et al. 2021 and references therein). Moreover, it was used for creating synthetic accelerograms based on numerical simulations or databases of real earthquake signals (Gatti and Clouteau 2020; Lehmann et al. 2022). ML was used as well in an attempt for earthquake prediction – the holy grail in seismology – showing that earthquake-like events in the laboratory could be predicted by identification by the machine of seismic precursor patterns (Johnson et al. 2021; Beroza et al. 2021; Laurenti et al. 2022; Ridzwan and Yusoff 2023). RL was also used for controlling earthquake-like events (Papachristos and Stefanou 2021) (see Chapter 4 of volume 2) that more exact mathematical theories show are controllable (Stefanou 2019; Gutiérrez-Oribio et al. 2022).

Given the striking advances of ML in self-driving cars, and object and face recognition (Balaban 2015), it is natural to expect ML methods to also find numerous applications in image processing in experimental geomechanics and geotechnics. *Convolutional neural networks* (CNNs) (see Chapter 5 of volume 2 for more details) are the basic ingredient of most ML techniques used in image and video processing, without forgetting more traditional compression techniques that are also considered as ML (e.g. PCA). Boukhtache et al. (2021) presents a review of digital image correlation (DIC) with deep learning. We also refer to some previous studies (Boukhtache et al. 2023; Cheng et al. 2023a; Duan et al. 2023) for some recent developments. Accuracy is one of the main issues for those methods in order to outperform the current state of the art (see ALERT Doctoral School 2022 (Ando et al. 2022)). Focusing on granular materials, (Cevallos et al. 2023) propose a CNN to automatically distinguish properly segmented digital grains with up to 90% accuracy, while (Cheng et al. 2023b) present an ML-based strategy to estimate the contact force chains of uniformly sized spherical granular materials using particle kinematics and inter-particle contact evolution data measured by X-ray micro-tomography.

1.4. Libraries for ML

Today, many libraries exist for ML (see https://en.wikipedia.org/wiki/Machine_learning#Software for an updated list). Most of the existing libraries provide a Python interface.

Some general purpose libraries for ML are Numpy (Harris et al. 2020), Pandas (McKinney 2010), SciPy (Virtanen et al. 2020) and Scikit-learn (Pedregosa et al. 2011). For ANNs, TensorFlow (Abadi et al. 2015) and PyTorch (Paszke et al. 2019) are equally popular at present (see Figure 1.5). Both offer parallelization and GPU support for training large ANN and handling large collections of data. The examples presented in these volumes make use of PyTorch (see Chapter 5 of volume 1 for more details). As always, the best library is the library that we know the best, provided that it is open source and it allows us to do what we want!

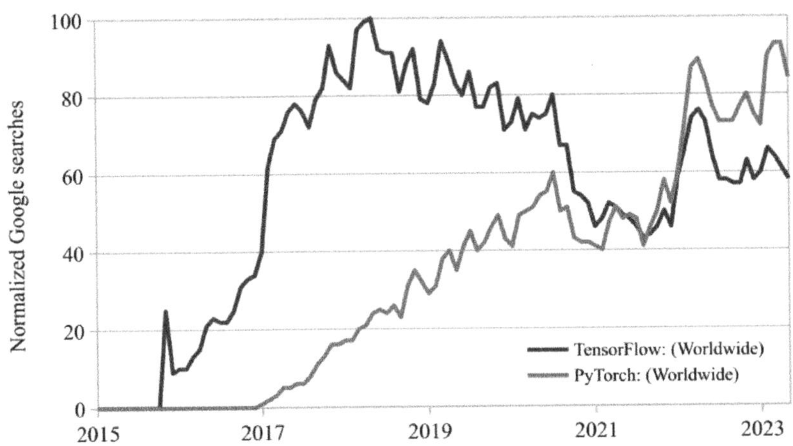

Figure 1.5. *Normalized Google searches over time for TensorFlow and PyTorch ANN libraries*

1.5. Bias in ML and limitations

The success of ML algorithms in performing complicated and complex tasks makes them convenient tools for many applications.

However, we have to be aware of their limitations. In other words, we have to use the right screwdriver for the right screw! In section 1.2.3 (see Figure 1.3), we showed how the choice of the ML method and the quality of the data can lead to biased predictions, even in a very simple problem.

Applications of ML algorithms have demonstrated gender (Kelion 2019), racial (Dressel and Farid 2018; Perkowitz 2021), hiring (Votto et al. 2021) and other biases (Varsha 2023). ML methods are algorithms which should be used with care and knowledge of the underlying limitations. This is not different from the application of other methods. For instance, we know how wrong finite element predictions can be when the appropriate finite elements are not used, when convergence analyses are not performed or when we want to model a softening material without regularizing the underlying mathematical problem. The inappropriate use of specific methods has traditionally led to spectacular failures with uncountable casualties and economic loss. ML methods will not be an exception, unless we understand them better and use them with caution.

As far as it concerns ML and geomechanics, luckily, we have at our disposal established and undeniable principles that have to be respected in any application. Conservation principles, the laws of thermodynamics and other physics at various spatiotemporal scales should be incorporated into the ML algorithms in order to assure adequate and safe predictions. Physics-based approaches in ML (see Chapters 1, 2 and 3 of volume 2) gain more and more attention from the scientific community and can become the natural environment for marrying the established know-how of decades of research in geomechanics with ML. This could give fresh ideas and an opportunity to push further the current state of the art in our fields. The incorporation of physics in ML could eventually inspire new ML methods in other domains too, which today suffer from inevitable bias (see social sciences where data are always limited).

Another limitation of most of the available ML methods today is their greediness for data. We do not need thousand of photos to teach a child what a car is! With ML though we need tons of data! In many applications, there is an abundance of data (see, for instance, the data that are produced, but not saved, at each increment at each Gauss point during a nonlinear finite element analysis in a geomechanics problem). In some other applications though, data may not be enough (see, for example, X-rays of thousand of specimens; Tengattini et al. 2020).

Noise in the data, overfitting and underfitting are some other points that we have to pay attention to. Data with a lot of noise can make the learning process hard and render the predictions unreliable and of poor generalization (see also Chapters 1–3 of volume 2). Data preparation to assure good quality is of paramount importance then.

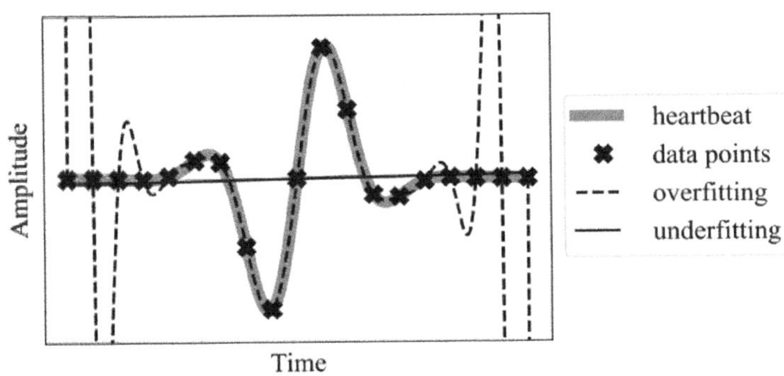

Figure 1.6. *Data points of a synthetic heartbeat, interpolated by two polynomial functions. The high-degree polynomial exactly fits the data points, but overfits the signal. The low-degree polynomial poorly fits the data points and underfits the signal. Both regressions are not acceptable, because they either predict a superhuman heartbeat or a dead person!*

The choice of a ML method allowing to fit data in a very high dimensional space can lead to overfitting. In this case, the training data can be very well represented, but predictions for unseen data can be far off. The opposite happens with underfitting, where the space for fitting is too low for identifying and reproducing the inherent patterns in the data. A classical example of overfitting and underfitting in polynomial regression is shown in Figure 1.6. Thankfully, there are ways for estimating in practice how well an ML model performs. This is mainly achieved by splitting the training set into training, validation and test data. *Regularization* methods can also help in avoiding this problem. For more details, we refer to Chapters 2, 4 and 5 of volume 1.

1.6. What to expect from these volumes?

These volumes aim to explain what ML is, what its main methods are and how they can be used for solving problems in (geo-)mechanics. Most of the chapters were written to provide a pedagogical introduction to the most important methods in ML and the fundamental notions behind them.

It is not possible to cover all the available ML methods in the existing literature and, without any doubt, many important methods were inevitably left out. For instance, we will not discuss genetic algorithms, PCA and related methods, particle swarm optimization, fuzzy logic algorithms, ML methods based on control theory and many others. We hope, however, to have provided a good selection of ML methods for an introductory course.

We expect the material presented in these volumes to help the readers to:

– demystify and understand what ML is;

– access the fundamental notions of the most important ML methods;

– use ML in simple examples, get aware of pitfalls and understand the need for physics-based ML methods for solving problems in (geo-)mechanics.

These volumes are addressed to undergraduate and graduate level. The minimum requirements for accessing them are:

– knowledge of Python programming language[1];

– basic concepts in mathematics (calculus and elements of differential calculus and numerical analysis);

1 For learning Python, we propose the book *Python Crash Course: A Hands-On, Project-Based Introduction to Programming* by Eric Matthes and/or many excellent tutorials that can be found on the internet.

– nice problems in mind that could combine ML and geomechanics!

1.7. Acknowledgments

The author would like to acknowledge the support of the European Research Council (ERC) under the European Union's Horizon2020 research and innovation program (Grant agreement no. 757848 CoQuake).

1.8. References

Abadi, M., Agarwal, A., Barham, P., Brevdo, E., Chen, Z., Citro, C., Corrado, G.S., Davis, A., Dean, J., Devin, M. et al. (2015). TensorFlow: Large-scale machine learning on heterogeneous systems [Online]. Available at: https://www.tensorflow.org/.

Allis, L.V. (1994). Searching for solutions in games and artificial intelligence. Doctoral Thesis, Maastricht University.

Ando, E., Marks, B., Hurley, R., Dijksman, J.A. (2022). Advanced experimental geomechanics. Lecture Notes, The Alliance of Laboratories in Europe for Education, Research and Technology [Online]. Available at: https://www.alertgeomaterials.eu/data/school/2022/2022_ALERT_school.pdf.

Baghbani, A., Choudhury, T., Costa, S., Reiner, J. (2022). Application of artificial intelligence in geotechnical engineering: A state-of-the-art review. *Earth-Science Reviews* 228, 103991 [Online]. Available at: https://linkinghub.elsevier.com/retrieve/pii/S0012825222000757.

Balaban, S. (2015). Deep learning and face recognition: The state of the art. *Biometric and Surveillance Technology for Human and Activity Identification XII*, Baltimore, MD, 94570B [Online]. Available at: http://proceedings.spiedigitallibrary.org/proceeding.aspx?doi=10.1117/12.2181526.

Becker, S., Devijver, E., Molinier, R., Jakse, N. (2022). Unsupervised topological learning approach of crystal nucleation. *Scientific Reports* 12(1), 3195 [Online]. Available at: https://www.nature.com/articles/s41598-022-06963-5.

Beroza, G.C., Segou, M., Mostafa Mousavi, S. (2021). Machine learning and earthquake forecasting—next steps. *Nature Communications* 12(1), 4761 [Online]. Available at: https://www.nature.com/articles/s41467-021-24952-6.

Boukhtache, S., Abdelouahab, K., Berry, F., Blaysat, B., Grédiac, M., Sur, F. (2021). When deep learning meets digital image correlation. *Optics and Lasers in Engineering*, 136, 106308 [Online]. Available at: https://linkinghub.elsevier.com/retrieve/pii/S0143816620306588.

Boukhtache, S., Abdelouahab, K., Bahou, A., Berry, F., Blaysat, B., Grédiac, M., Sur, F. (2023). A lightweight convolutional neural network as an alternative to DIC to measure in-plane displacement fields. *Optics and Lasers in Engineering*, 161, 107367 [Online]. Available at: https://linkinghub.elsevier.com/retrieve/pii/S0143816622004183.

Cevallos, S.B., Jerves, A.X., Mital, U., Medina, D.A., Quinteros, V.S., Mulas, M., Torgersrud, N. (2023). Towards a more accurate characterization of granular media 2.0: Involving AI in the process. *Computers and Geotechnics*, 160, 105510 [Online]. Available at: https://linkinghub.elsevier.com/retrieve/pii/S0266352X23002677.

Cheng, X., Zhou, S., Xing, T., Zhu, Y., Ma, S. (2023a). Solving digital image correlation with neural networks constrained by strain-displacement relations. *Optics Express*, 31(3), 3865 [Online]. Available at: https://opg.optica.org/abstract.cfm?URI=oe-31-3-3865.

Cheng, Z., Wang, J., Xiong, W. (2023b). A machine learning-based strategy for experimentally estimating force chains of granular materials using X-ray micro-tomography. *Géotechnique*, 1–13 [Online]. Available at: https://www.icevirtuallibrary.com/doi/10.1680/jgeot.21.00281.

Dressel, J. and Farid, H. (2018). The accuracy, fairness, and limits of predicting recidivism. *Science Advances*, 4(1), eaao5580 [Online]. Available at: https://www.science.org/doi/10.1126/sciadv.aao5580.

Duan, X., Xu, H., Dong, R., Lin, F., Huang, J. (2023). Digital image correlation based on convolutional neural networks. *Optics and Lasers in Engineering* 160, 107234 [Online]. Available at: https://linkinghub.elsevier.com/retrieve/pii/S0143816622002871.

Eijnden, A.P.V.D., Bésuelle, P., Collin, F., Chambon, R, Desrues, J. (2017). Modeling the strain localization around an underground gallery with a hydro-mechanical double scale model; effect of anisotropy. *Computers and Geotechnics*, 85, 384–400. doi: 10.1016/j.compgeo.2016.08.006.

Feyel, F. (2003). A multilevel finite element method (FE2) to describe the response of highly non-linear structures using generalized continua. *Computer Methods in Applied Mechanics and Engineering* 192(28–30), 3233–3244.

Gatti, F. and Clouteau, D. (2020). Towards blending physics-based numerical simulations and seismic databases using generative adversarial network. *Computer Methods in Applied Mechanics and Engineering* 372, 113421 [Online]. Available at: https://linkinghub.elsevier.com/ retrieve/pii/S004578252030606X.

Géron, A. (2019). *Hands-On Machine Learning with Scikit-Learn, Keras, and TensorFlow*. O'Reilly Media, Sebastopol, CA.

Ghaboussi, J., Garrett, J.H., Wu, X. (1991). Knowledge based modeling of material behavior with neural networks. *Journal of Engineering Mechanics*, 117(1), 132–153.

Guo, Z., Bai, R., Lei, Z., Jiang, H., Liu, D., Zou, J., Yan, C. (2021). CPINet: Parameter identification of path-dependent constitutive model with automatic denoising based on CNN-LSTM. *European Journal of Mechanics – A/Solids*, 90, 104327 [Online]. Available at: https://linkinghub.elsevier.com/retrieve/pii/S099775382100098X.

Gutiérrez-Oribio, D., Tzortzopoulos, G., Stefanou, I., Plestan, F. (2022). Earthquake control: An emerging application for robust control. Theory and experimental tests. *IEEE Transactions on Control Systems Technology*, 31, 1747–1761.

Harris, C.R., Millman, K.J., Walt, S.J.V. D., Gommers, R., Virtanen, P., Cournapeau, D., Wieser, E., Taylor, J., Berg, S., Smith, N.J. et al. (2020). Array programming with NumPy. *Nature*, 585(7825), 357–362. doi: 10.1038/s41586-020-2649-2.

Heider, Y., Wang, K., Sun, W. (2020). SO(3)-invariance of informed-graph-based deep neural network for anisotropic elastoplastic materials. *Computer Methods in Applied Mechanics and Engineering* 363, 112875.

Hernandez, Q., Badías, A., González, D., Chinesta, F., Cueto, E. (2021). Deep learning of thermodynamics-aware reduced-order models from data. *Computer Methods in Applied Mechanics and Engineering*, 379, 113763 [Online]. Available at: https://linkinghub.elsevier.com/retrieve/pii/S0045782521000992.

Huang, D.Z., Xu, K., Farhat, C., Darve, E. (2020). Learning constitutive relations from indirect observations using deep neural networks. *Journal of Computational Physics*, 416, 109491.

Johnson, P.A., Rouet-Leduc, B., Pyrak-Nolte, L.J., Beroza, G.C., Marone, C.J., Hulbert, C., Howard, A., Singer, P., Gordeev, D., Karaflos, D. et al. (2021). Laboratory earthquake forecasting: A machine learning competition. *Proceedings of the National Academy of Sciences*, 118(5), e2011362118 [Online]. Available at: https://pnas.org/doi/full/10.1073/pnas.2011362118.

Karapiperis, K., Stainier, L., Ortiz, M. and Andrade, J. (2020). Data-driven multiscale modeling in mechanics. *Journal of the Mechanics and Physics of Solids*, 147, 104239. doi: 10.1016/j.jmps.2020.104239.

Karapiperis, K., Ortiz, M., Andrade, J. (2021). Data-driven nonlocal mechanics: Discovering the internal length scales of materials. *Computer Methods in Applied Mechanics and Engineering*, 386, 114039. doi: 10.1016/j.cma.2021.114039.

Karniadakis, G.E., Kevrekidis, I.G., Lu, L., Perdikaris, P., Wang, S., Yang, L. (2021). Physics-informed machine learning. *Nature Reviews Physics*, 3(6), 422–440 [Online]. Available at: https://www.nature.com/articles/s42254-021-00314-5.

Kelion, L. (2019). Apple's 'sexist' credit card investigated by US regulator. *BBC* [Online]. Available at: https://www.bbc.com/news/business-50365609.

Kirchdoerfer, T. and Ortiz, M. (2016). Data-driven computational mechanics. *Computer Methods in Applied Mechanics and Engineering*, 304, 81–101. doi: 10.1016/j.cma.2016.02.001.

Kirchdoerfer, T. and Ortiz, M. (2018). Data-driven computing in dynamics. *International Journal for Numerical Methods in Engineering*, 113(11), 1697–1710.

Laurenti, L., Tinti, E., Galasso, F., Franco, L., Marone, C. (2022). Deep learning for laboratory earthquake prediction and autoregressive forecasting of fault zone stress. *Earth and Planetary Science Letters*, 598, 117825 [Online]. Available at: https://linkinghub.elsevier.com/retrieve/pii/S0012821X22004617.

Lefik, M. and Schrefler, B.A. (2003). Artificial neural network as an incremental non-linear constitutive model for a finite element code. *Computer Methods in Applied Mechanics and Engineering*, 192(28–30), 3265–3283.

Lehmann, F., Gatti, F., Bertin, M., Clouteau, D. (2022). Machine learning opportunities to conduct high-fidelity earthquake simulations in multi-scale heterogeneous geology. *Frontiers in Earth Science*, 10, 1029160 [Online]. Available at: https://www.frontiersin.org/articles/10.3389/feart.2022.1029160/full.

Liu, Z. and Wu, C.T. (2019). Exploring the 3D architectures of deep material network in data-driven multiscale mechanics. *Journal of the Mechanics and Physics of Solids*, 127, 20–46. arXiv: 1901.04832.

Lloberas Valls, O., Raschi Schaw, M., Huespe, A.E., Oliver Olivella, X. (2019). Reduced finite element square techniques (RFE2): Towards industrial multiscale FE software. In *COMPLAS 2019: XV International Conference on Computational Plasticity: Fundamentals and Applications*. International Centre for Numerical Methods in Engineering (CIMNE).

Marques, A., Pereira, A., Ribeiro, B., Prates, P.A. (2022). On the identification of material constitutive model parameters using machine learning algorithms. *Key Engineering Materials*, 926, 2146–2153 [Online]. Available at: https://www.scientific.net/KEM.926.2146.

Masi, F. and Stefanou, I. (2022). Multiscale modeling of inelastic materials with thermodynamics-based artificial neural networks (TANN). *Computer Methods in Applied Mechanics and Engineering*, 398, 115190 [Online]. Available at: https://linkinghub.elsevier.com/retrieve/pii/S0045782522003450.

Masi, F. and Stefanou, I. (2023). Evolution TANN and the identification of internal variables and evolution equations in solid mechanics. *Journal of the Mechanics and Physics of Solids*, 105245 [Online]. Available at: https://linkinghub.elsevier.com/retrieve/pii/S0022509623000492.

Masi, F., Stefanou, I., Vannucci, P., Maffi-Berthier, V. (2021). Thermodynamics-based artificial neural networks for constitutive modeling. *Journal of the Mechanics and Physics of Solids* 147, 104277 [Online]. Available at: http://arxiv.org/abs/2005.12183.

McKinney, W. (2010). Data structures for statistical computing in Python. In *Proceedings of the 9th Python in Science Conference*, Austin, TX.

Mitchell, T.M. (1997). *Machine Learning*. McGraw-Hill, New York.

Morand, L. and Helm, D. (2019). A mixture of experts approach to handle ambiguities in parameter identification problems in material modeling. *Computational Materials Science*, 167, 85–91 [Online]. Available at: https://linkinghub.elsevier.com/retrieve/pii/S0927025619302058.

Mozaffar, M., Bostanabad, R., Chen, W., Ehmann, K., Cao, J., Bessa, M.A. (2019). Deep learning predicts path-dependent plasticity. *Proceedings of the National Academy of Sciences* 116(52), 26414–26420.

Nguyen, T.K., Combe, G., Caillerie, D., Desrues, J. (2014). FEM × DEM modelling of cohesive granular materials: Numerical homogenisation and multi-scale simulations. *Acta Geophysica* 62(5), 1109–1126 [Online]. Available at: http://link.springer.com/10.2478/s11600-014-0228-3.

Nitka, M., Combe, G., Dascalu, C., Desrues, J. (2011). Two-scale modeling of granular materials: A DEM-FEM approach. *Granular Matter*, 13(3), 277–281.

Papachristos, E. and Stefanou, I. (2021). Controlling earthquake-like instabilities using artificial intelligence. *arXiv* [Online]. Available at: http://arxiv.org/abs/2104.13180.

Paszke, A., Gross, S., Massa, F., Lerer, A., Bradbury, J., Chanan, G., Killeen, T., Lin, Z., Gimelshein, N., Antiga, L. et al. (2019). PyTorch: An imperative style, high-performance deep learning library. In *Advances in Neural Information Processing Systems 32 (NeurIPS 2019)*. Curran Associates, Inc, Vancouver [Online]. Available at: http://papers.neurips.cc/paper/9015-pytorch-an-imperative-style-high-performance-deep-learning-library.pdf.

Pedregosa, F., Varoquaux, G., Gramfort, A., Michel, V., Thirion, B., Grisel, O., Blondel, M., Prettenhofer, P., Weiss, R., Dubourg, V. et al. (2011). Scikit-learn: Machine learning in Python. *Journal of Machine Learning Research*, 12, 2825–2830.

Peng, G.C.Y., Alber, M., Buganza Tepole, A., Cannon, W.R., De, S., Dura-Bernal, S., Garikipati, K., Karniadakis, G., Lytton, W.W., Perdikaris, P. et al. (2021). Multiscale modeling meets machine learning: What can we learn? *Archives of Computational Methods in Engineering*, 28(3), 1017–1037 [Online]. Available at: https://link.springer.com/10.1007/s11831-020-09405-5.

Perkowitz, S. (2021). The bias in the machine: Facial recognition technology and racial disparities. *MIT Case Studies in Social and Ethical Responsibilities of Computing*, February [Online]. Available at: https://mit-serc.pubpub.org/pub/bias-in-machine.

Pouragha, M., Eghbalian, M., Wan, R. (2020). A mote on applicability of artificial intelligence to constitutive modeling of geomaterials. *Journal of Machine Learning for Modeling and Computing*, 1(2), 157–170 [Online]. Available at: http://www.dl.begellhouse.com/journals/558048804a15188a,2cbcbe11139f18e5,478af42571e4a7ae.html.

Raissi, M., Perdikaris, P., Karniadakis, G.E. (2019). Physics-informed neural networks: A deep learning framework for solving forward and inverse problems involving nonlinear partial differential equations. *Journal of Computational Physics*, 378, 686–707. doi: 10.1016/j.jcp.2018.10.045.

Ridzwan, N.S.M. and Yusoff, S.H.M. (2023). Machine learning for earthquake prediction: A review (2017–2021). *Earth Science Informatics*, 16(2), 1133–1149 [Online]. Available at: https://link.springer. com/10.1007/s12145-023-00991-z.

Rocha, I., Kerfriden, P., Van Der Meer, F. (2021). On-the-fly construction of surrogate constitutive models for concurrent multiscale mechanical analysis through probabilistic machine learning. *Journal of Computational Physics: X*, 9, 100083 [Online]. Available at: https://linkinghub.elsevier.com/retrieve/pii/S2590055220300354.

Schulte, R., Karca, C., Ostwald, R., Menzel, A. (2023). Machine learning-assisted parameter identification for constitutive models based on concatenated loading path sequences. *European Journal of Mechanics – A/Solids*, 98, 104854 [Online]. Available at: https://linkinghub.elsevier.com/retrieve/pii/S0997753822002844.

Shannon, C.E. (1950). XXII. Programming a computer for playing chess. *The London, Edinburgh, and Dublin Philosophical Magazine and Journal of Science*, 41(314), 256–275 [Online]. Available at: http://www.tandfonline.com/doi/abs/10.1080/14786445008521796.

Silver, D., Huang, A., Maddison, C.J., Guez, A., Sifre, L., Van Den Driessche, G., Schrittwieser, J., Antonoglou, I., Panneershelvam, V., Lanctot, M. et al. (2016). Mastering the game of Go with deep neural networks and tree search. *Nature*, 529(7587), 484–489 [Online]. Available at: https://www.nature.com/articles/nature16961.

Sircar, A., Yadav, K., Rayavarapu, K., Bist, N., Oza, H. (2021). Application of machine learning and artificial intelligence in oil and gas industry. *Petroleum Research*, 6(4), 379–391 [Online]. Available at: https://linkinghub.elsevier.com/retrieve/pii/S2096249521000429.

Stefanou, I. (2019). Controlling anthropogenic and natural seismicity: Insights from active stabilization of the Spring–Slider model. *Journal of Geophysical Research: Solid Earth*, 124(8), 8786–8802 [Online]. Available at: https://onlinelibrary.wiley.com/doi/abs/10.1029/2019JB017847.

Sun, X., Bahmani, B., Vlassis, N.N., Sun, W., Xu, Y. (2022). Data-driven discovery of interpretable causal relations for deep learning material laws with uncertainty propagation. *Granular Matter*, 24, 1. doi: 10.1007/s10035-021-01137-y.

Tehrani, F.S., Calvello, M., Liu, Z., Zhang, L., Lacasse, S. (2022). Machine learning and landslide studies: Recent advances and applications. *Natural Hazards*, 114(2), 1197–1245 [Online]. Available at: https://link.springer.com/10.1007/s11069-022-05423-7.

Tengattini, A., Lenoir, N., Andò, E., Giroud, B., Atkins, D., Beaucour, J., Viggiani, G. (2020). NeXT-Grenoble, the neutron and X-ray tomograph in Grenoble. *Nuclear Instruments and Methods in Physics Research Section A: Accelerators, Spectrometers, Detectors and Associated Equipment*, 968, 163939 [Online]. Available at: https://linkinghub.elsevier.com/retrieve/pii/S0168900220304198.

Tromp, J. and Farnebäck, G. (2007). Combinatorics of Go. In *Computers and Games*, Hutchison, D., Kanade, T., Kittler, J. et al. (eds). Springer, Berlin, Heidelberg [Online]. Available at: http://link.springer.com/10.1007/978-3-540-75538-8_8.

Varsha, P.S. (2023). How can we manage biases in artificial intelligence systems – A systematic literature review. *International Journal of Information Management Data Insights*, 3(1), 100165 [Online]. Available at: https://linkinghub.elsevier.com/retrieve/pii/S2667096823000125.

Virtanen, P., Gommers, R., Oliphant, T.E., Haberland, M., Reddy, T., Cournapeau, D., Burovski, E., Peterson, P., Weckesser, W., Bright, J. et al. (2020). SciPy 1.0: Fundamental algorithms for scientific computing in Python. *Nature Methods*, 17, 261–272.

Vlassis, N.N. and Sun, W. (2021). Component-based machine learning paradigm for discovering rate-dependent and pressure-sensitive level-set plasticity models. *Journal of Applied Mechanics*, 1–13 [Online]. Available at: https://asmedigitalcollection.asme.org/appliedmechanics/article/doi/10.1115/1.4052684/1121215/Component-based-machine-learning-paradigm-for.

Votto, A.M., Valecha, R., Najafirad, P., Rao, H.R. (2021). Artificial intelligence in tactical human resource management: A systematic literature review. *International Journal of Information Management Data Insights*, 1(2), 100047 [Online]. Available at: https://linkinghub.elsevier.com/retrieve/pii/S2667096821000409.

Wang, K., Sun, W., Du, Q. (2019). A cooperative game for automated learning of elasto-plasticity knowledge graphs and models with AI-guided experimentation. *Computational Mechanics*, 64(2), 467–499 [Online]. Available at: http://link.springer.com/10.1007/s00466-019-01723-1.

Wu, M., Xia, Z., Wang, J. (2023). Constitutive modelling of idealised granular materials using machine learning method. *Journal of Rock Mechanics and Geotechnical Engineering*, 15(4), 1038–1051 [Online]. Available at: https://linkinghub.elsevier.com/retrieve/pii/S1674775522001688.

Zhang, W. and Liu, Z. (2022). Editorial for machine learning in geotechnics. *Acta Geotechnica*, 17(4), 1017–1017 [Online]. Available at: https://link.springer.com/10.1007/s11440-022-01563-z.

Zhang, P., Yin, Z.-Y., Jin, Y.-F. (2021a). State-of-the-art review of machine learning applications in constitutive modeling of soils. *Archives of Computational Methods in Engineering*, 28(5), 3661–3686 [Online]. Available at: https://link.springer.com/10.1007/s11831-020-09524-z.

Zhang, W., Li, H., Li, Y., Liu, H., Chen, Y., Ding, X. (2021b). Application of deep learning algorithms in geotechnical engineering: A short critical review. *Artificial Intelligence Review*, 54(8), 5633–5673 [Online]. Available at: https://link.springer.com/10.1007/s10462-021-09967-1.

Zhang, N., Zhou, A., Jin, Y.-F., Yin, Z.-Y., Shen, S.-L. (2023). An enhanced deep learning method for accurate and robust modelling of soil stress–strain response. *Acta Geotechnica* [Online]. Available at: https://link.springer.com/10.1007/s11440-023-01813-8.

Zhou, X.-H., Han, J., Xiao, H. (2021). Learning nonlocal constitutive models with neural networks. *Computer Methods in Applied Mechanics and Engineering* 384, 113927 [Online]. Available at: https://linkinghub.elsevier.com/retrieve/pii/ S0045782521002644.

2

Introduction to Regression Methods

Filippo MASI
Sydney Centre in Geomechanics and Mining Materials, School of Civil Engineering, University of Sydney, Australia

Regression is one of the fundamental pillars of supervised machine learning (ML). In this chapter, we uncover the essential concepts in regression analysis and methods, by providing (hands-on) practical examples, designed for graduate students and researchers seeking to gain a solid understanding.

In particular, we first delve into linear regression methods, exploring both closed-form solutions and the general optimization framework provided by the gradient descent (GD). Then, we introduce the necessary pre- and post-processing steps for building and testing models: feature scaling, hold-out and cross-validation.

Moving on, we focus on nonlinear regression methods and, in particular, polynomial regression. Through extensive examples, we

For a color version of all the figures in this chapter, see www.iste.co.uk/stefanou/machine1.zip.

Machine Learning in Geomechanics 1,
coordinated by Ioannis STEFANOU and Félix DARVE.

then demonstrate the benefits associated with regularization techniques (LASSO, Ridge and Elastic Net) and the possibility of constructing interpretable and parsimonious models.

Lastly, we introduce Bayesian approaches with focus on linear and Gaussian process (GP) regression. The advantages and drawbacks of the latter are compared with ordinary regression methods.

2.1. Introduction

Regression is a statistical technique used to explore and quantify the relationship between dependent variables (often called outcomes) and independent variables (often called predictors). In other words, regression involves fitting a model to some data, under some error function. Two distinct purposes characterize regression. The first one, and the primary, is to identify a model that can be used for predictions and forecasting, while the second purpose is to infer causal relationships between independent and dependent variables under certain circumstances.

To grasp the main core of regression, consider a scenario where we seek to understand the influence of relative density and effective confining pressure on the shear strength and critical state behavior of a sand sample. Figure 2.1(a) presents 25 drained monotonic triaxial compression tests of Karlsruhe fine sand with different initial relative densities I_{D0} and effective confining pressures (Wichtmann and Triantafyllidis 2016). In this framework, regression allows us to identify the intrinsic relationship between independent variables $\mathbf{x}$ (i.e. relative density and effective pressure) and a dependent variable y (shear strength or critical state line) by assuming:

$$y = f_{\boldsymbol{\theta}}(\mathbf{x}) + \epsilon \qquad [2.1]$$

where $f_{\boldsymbol{\theta}}$ represents a function, with $\boldsymbol{\theta}$ denoting its parameters (unknown), and ϵ is an error term.[1]

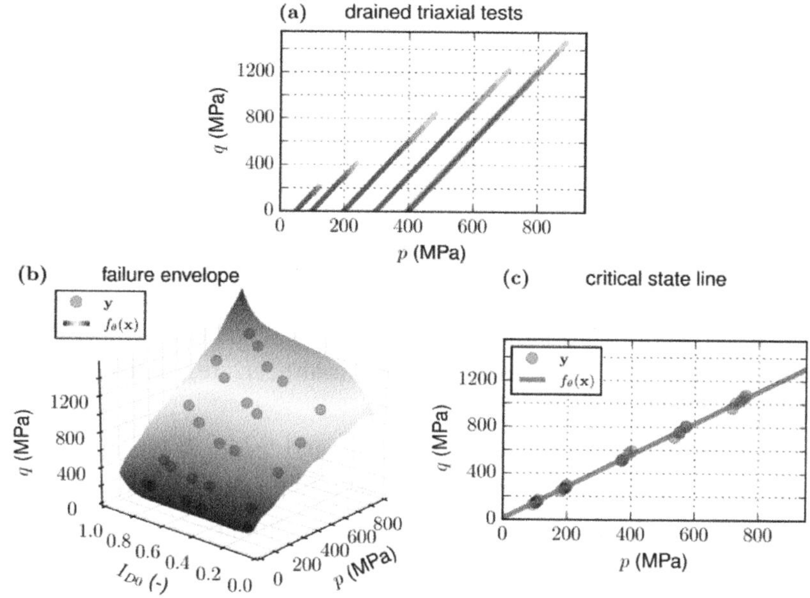

Figure 2.1. *Drained monotonic triaxial compression tests of Karlsruhe fine sand with different initial relative densities I_{D0} and effective confining pressures (Wichtmann and Triantafyllidis 2016) (a) Nonlinear regression of the failure envelope locus of peak deviatoric (b) and critical state line (c) using a basis of polynomial and trigonometric functions*

With this objective in mind, we continue by formulating an objective function that measures the difference between the predictions of the model [2.1] and the values of the dependent variable. Through the minimization of the latter, we can determine an optimal fit and thus identify the parameters $\boldsymbol{\theta}$. In so doing, we obtain two models ($f_{\boldsymbol{\theta}}(\mathbf{x})$), which are able to generate predictions of the failure envelope and critical state, as depicted in Figures 2.1(b) and (c). This is the core

1 The error term ϵ is an irreducible source of error. For instance, it may arise whenever there exist variables other than $\mathbf{x}$ – and independent of $\mathbf{x}$ – that have some not negligible effect on the dependent variable y.

process of regression and enables us to uncover relationships between variables (data), make predictions and gain insights from data.

This chapter provides the tools necessary for carrying out such analyses by means of a comprehensive introduction to regression methods using a "hands-on approach", that is, providing essential theoretical aspects and prioritizing understanding through concrete examples. After studying this chapter, we hope that the reader will be able to:

1) understand the fundamental notions related to regression methods and GD;

2) solve a regression problem and select the best (class of) method(s) depending on the particular task (linear, nonlinear, regularized or Bayesian regression);

3) understand the importance of validating and testing a regression model (and in general any ML model) and how to do so;

4) grasp the importance of regularization techniques to uncover parsimonious and simple models from data sets;

5) understand the benefits and drawbacks of Bayesian approaches as compared to ordinary regression methods.

All codes used in this chapter are available on ALERT Geomaterials GitHub (https://github.com/alert-geomaterials/2023-doctoral-school).

As it follows, scalar values and functions are represented in *lowercase italic* font, while **lowercase bold** font denotes vectors and **uppercase bold** font denotes matrices.

2.2. Linear regression

In linear regression (Mandel 1982), the idea is to select a linear function $f_{\boldsymbol{\theta}}(\cdot)$, in equation [2.1], whose parameters $\boldsymbol{\theta}$ are identified to fit data by means of the minimization of some error metric between the predictions, that is, the outputs of the function, and the dependent variables. As we will see, linear regression can be formulated as a simple solution of a linear system.

To fix the ideas, let us start by considering a *training* data set, with n data points or *snapshots*, composed of p dependent variables $\mathbf{Y}$ and m independent variables $\mathbf{X}$:

$$\mathbf{Y} \equiv \begin{bmatrix} | & | & | & | \\ \mathbf{y}^{(1)} & \mathbf{y}^{(2)} & \cdots & \mathbf{y}^{(n)} \\ | & | & | & | \end{bmatrix}, \qquad \mathbf{X} \equiv \begin{bmatrix} | & | & | & | \\ \mathbf{x}^{(1)} & \mathbf{x}^{(2)} & \cdots & \mathbf{x}^{(n)} \\ | & | & | & | \end{bmatrix}, \qquad [2.2]$$

where $\mathbf{y}^{(k)}$ and $\mathbf{x}^{(k)}$ are the vectors collecting the p dependent variables and the m independent variables, respectively, at the kth snapshot of the data set.

In the following, we will consider for simplicity one-dimensional dependent variables, that is, $p = 1$, and we will refer to the vector collecting the snapshots of the dependent variable as $\mathbf{y}$. However, the entire framework developed herein can be easily formulated in higher dimensions.

To find a best fit line through the training data points, we thus assume a linear model of the form:

$$\hat{y} = \theta_0 + \theta_1 x_1 + \theta_2 x_2 + \cdots + \theta_m x_m, \qquad [2.3]$$

where $\hat{y}$ is the value predicted by the model, x_i is the ith independent variable value and θ_i is the ith model parameter – including the *bias* term, θ_0, and the model *weights*, $\theta_1, \theta_2, \cdots, \theta_m$. The above equation can be formulated more concisely as:

$$\hat{y} = f_{\boldsymbol{\theta}}(\mathbf{x}) \equiv \boldsymbol{\theta}^{\mathrm{T}}\mathbf{x}^*, \qquad [2.4]$$

where $\boldsymbol{\theta} = \left[\theta_0, \theta_1, \theta_2, \cdots, \theta_m\right]^{\mathrm{T}}$ is the vector collecting the model parameters, $\mathbf{x}^* = \left[x_0, x_1, x_2, \cdots, x_m\right]^{\mathrm{T}}$ is the augmented feature vector with $x_0 = 1$, $f_{\boldsymbol{\theta}}(\cdot)$ is often called *hypothesis function* – a linear function, in this case – and the superscript T denotes the transpose operator. To simplify notation, in the following, we write interchangeably $\mathbf{x}$ for both $\mathbf{x}$ and $\mathbf{x}^*$.

The next step is to identify the parameters $\boldsymbol{\theta}$ such that the linear regression model, $f_{\boldsymbol{\theta}}(\mathbf{x})$, best fits the data set $\{\mathbf{X}, \mathbf{y}\}$. This operation is called *training*. Training identifies the parameter values that optimize a measure of the *goodness of fit*, that is, how well (or poorly) a model fits the data. Various objective functions, often referred to as loss functions, can be adopted and their choice strongly determines the subsequent model. Two standard error metrics of the hypothesis $f_{\boldsymbol{\theta}}(\cdot)$ on a set $\mathbf{X}$ are often considered, which are associated with the ℓ_1 and ℓ_2 norms, respectively, defined as the mean absolute error (MAE) and the root mean square error (RMSE):

$$E_1(\mathbf{X}, f_{\boldsymbol{\theta}}) \equiv \frac{1}{n}\sum_{k=1}^{n}\left|f_{\boldsymbol{\theta}}(\mathbf{x}^{(k)}) - y^{(k)}\right| \qquad [2.5a]$$

$$E_2(\mathbf{X}, f_{\boldsymbol{\theta}}) \equiv \left(\frac{1}{n}\sum_{k=1}^{n}\left(f_{\boldsymbol{\theta}}(\mathbf{x}^{(k)}) - y^{(k)}\right)^2\right)^{1/2}. \qquad [2.5b]$$

One can also broadly define the error based on the ℓ_r norm, and namely:

$$E_r(\mathbf{X}, f_{\boldsymbol{\theta}}) \equiv \left(\frac{1}{n}\sum_{k=1}^{n}\left|f_{\boldsymbol{\theta}}(\mathbf{x}^{(k)}) - y^{(k)}\right|^r\right)^{1/r}. \qquad [2.6]$$

The higher the norm index r, the more the error metrics focuses on large values and neglects small ones, thus the best fit model intrinsically depends on r. In most cases, the differences between models based on different norms are small. However, when there are outliers[2] in the data, the choice of the norm can have a significant impact: for instance, the RMSE is more sensitive to outliers than the MAE.

In linear regression, the most common choice is the root mean square error, but in practice, it is simpler to minimize the mean squared error (MSE) than the RMSE, and it leads to the same result

2 An outlier is an observation that lies an abnormal distance from the other values present in a data set.

(because the value of $\boldsymbol{\theta}$ that minimizes a function also minimizes its square root). The MSE of a linear regression hypothesis $f_{\boldsymbol{\theta}}$ on a set $\mathbf{X}$ is given by:

$$\text{MSE}\left(\mathbf{X}, f_{\boldsymbol{\theta}}\right) \equiv E_2^2\left(\mathbf{X}, f_{\boldsymbol{\theta}}\right) = \frac{1}{n}\sum_{k=1}^{n}\left(\boldsymbol{\theta}^{\mathrm{T}}\mathbf{x}^{(k)} - y^{(k)}\right)^2. \qquad [2.7]$$

Once the loss function is defined, training involves the identification of the parameters that minimize that particular loss. This requires differentiation with respect to $\boldsymbol{\theta}$ to identify the value of the latter such that a minimum of the error occurs, that is, find those $\boldsymbol{\theta}$ for which $\partial\text{MSE}\left(\mathbf{X}, f_{\boldsymbol{\theta}}\right)/\partial\boldsymbol{\theta} = 0$. Note that although a zero derivative denotes either a minimum or a maximum, we know this must be a minimum of the error since there is no maximum error, that is, we can always find a model that has a larger error.

Following the above procedure, the values of the parameters for which the error is minimum ($\hat{\boldsymbol{\theta}}$) are the solution of the following linear system of equations:

$$\mathbf{X}^{\mathrm{T}}\mathbf{X}\hat{\boldsymbol{\theta}}^{\mathrm{T}} = \mathbf{X}^{\mathrm{T}}\mathbf{y}. \qquad [2.8]$$

If the matrix $\mathbf{X}^{\mathrm{T}}\mathbf{X}$ is square and invertible (i.e. it has non-zero determinant), then there exists a unique solution $\hat{\boldsymbol{\theta}}$, which is given by the *normal equation*:

$$\hat{\boldsymbol{\theta}}^{\mathrm{T}} = \left(\mathbf{X}^{\mathrm{T}}\mathbf{X}\right)^{-1}\mathbf{X}^{\mathrm{T}}\mathbf{y}. \qquad [2.9]$$

However, when $\mathbf{X}^{\mathrm{T}}\mathbf{X}$ is singular, the normal equation does not hold. An alternative way to solve equation [2.4] consists of using the Moose–Penrose pseudoinverse[3] $\mathbf{X}^{+}$ of $\mathbf{X}$ to obtain the general solution:

---

3 The pseudoinverse is computed using the matrix factorization technique called singular value decomposition that can decompose the independent variables matrix $\mathbf{X}$ into the multiplication of three matrices $\mathbf{U}\boldsymbol{\Sigma}\mathbf{V}^{\mathrm{T}}$ (assuming $\mathbf{X}$ is a real-valued matrix). The pseudoinverse is computed as $\mathbf{X} = \mathbf{V}\boldsymbol{\Sigma}^{+}\mathbf{U}^{\mathrm{T}}$. To compute the matrix $\boldsymbol{\Sigma}^{+}$, the algorithm takes $\boldsymbol{\Sigma}$ and sets to zero all values smaller than a tiny threshold value, then it replaces all the non-zero values with their inverse, and finally it transposes the resulting matrix.

$$\hat{\theta} = \mathbf{X}^{+}\mathbf{y}, \tag{2.10}$$

obtained by leveraging the commutative property $(\mathbf{X}^{+})^{\mathrm{T}} = (\mathbf{X}^{\mathrm{T}})^{+}$.

We now have all the necessary ingredients for building a linear regression model: it is time to move to a simple example.

2.2.1. *Example*

Let us consider fitting a data set, as shown in Figure 2.2, generated by the function:

$$y = \alpha_2 \exp\left(\alpha_1 x + \mathcal{U}\left(-1, 1\right)\right), \quad \alpha_1 = 1,\ \alpha_2 = e, \tag{2.11}$$

where $\mathcal{U}(-1,1)$ is a uniformly distributed random variable that lies between -1 and 1. The snippet hereinafter generates our data set:

```
import numpy as np
np.random.seed(42)
n_snapshots = 200 # set number of snapshots: n
a_1 = 1.; a_2 = np.exp(1) # set coefficients
noise = np.random.uniform(-1,1,(n_snapshots,1)) # uniform noise
X = np.random.uniform(0,4.,(n_snapshots,1))
                                        # independent variable X
y = a_2*np.exp(a_1*X+noise) # dependent variable y
```

At this point, we can use the normal equation [2.9] and compute the values $\hat{\theta}$ using NumPy's linear algebra module (`np.linalg`) to calculate the inverse of a matrix and the `dot()` method for matrix multiplication:

```
X_p = np.c_[np.ones((n_snapshots, 1)), X]
                                # add x0 = 1 to each instance
theta_best = np.linalg.inv(X_p.T.dot(X_p)).dot(X_p.T).dot(y)
print(theta_best) # parameters
array([[-16.93873601],
       [ 28.05702557]])
```

The normal equation gives a linear model of the form $\hat{y} \approx -16.9 + 28.0x$. Let us check if the model correctly fits the data set, making predictions as follows:

```
X_new = np.array([[0],[4]])
X_b = np.c_[np.ones((2,1)), X_new] # add x0 = 1 to each instance
y_predict = X_b.dot(theta_best)
```

Finally, we can plot the predictions (y_predict) with the training data set (see Figure 2.2). As one could have imagined from the very beginning, the linear regression model poorly fits the nonlinear data set.

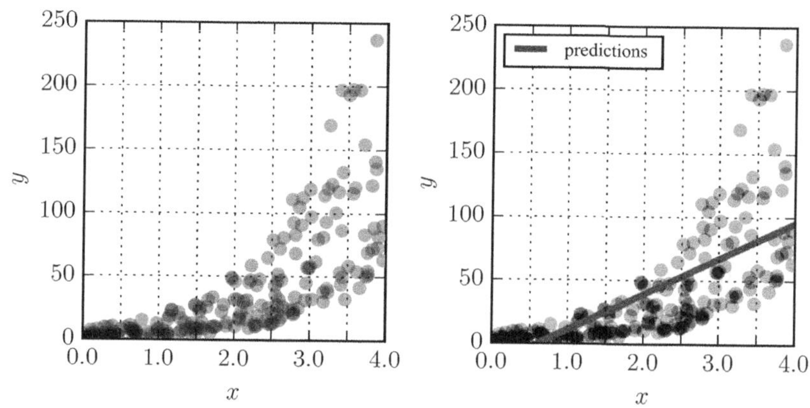

Figure 2.2. *Linear regression: an example and a failure (!). Randomly generated data set (left) –* $y = \alpha_2 e^{(\alpha_1 x + \mathcal{U}(-1,1))}$ *– and linear regression model predictions (right)*

But, as in most cases, data set preparation and transformation may help us in improving the model predictions. In this particular case, one has to simply perform a change of coordinates, defining $y' = \ln(y)$ and $\alpha_2' = \ln(\alpha_2)$. Such a transformation (*magically*) performs the linearization of the original data set, which from [2.11] now becomes:

$$y' = \alpha_2' + \alpha_1 x + \mathcal{U}(-1,1), \quad \alpha_1 = 1,\ \alpha_2' = 1, \qquad [2.12]$$

Let us try to apply the following changes to the original data set, repeating the linear regression fit, and plotting the predictions:

```
y_prime = np.log(y) # change of coordinate, y' = ln(y)
X_b = np.c_[np.ones((n_snapshots, 1)), X]
                              # add x0 = 1 to each instance
theta_best = np.linalg.inv(X_b.T.dot(X_b)).dot(X_b.T).dot(y_prime)
X_new = np.array([[0],[4]])
X_b = np.c_[np.ones((2,1)), X_new] # add x0 = 1 to each instance
y_predict = X_b.dot(theta_best) # parameters
print(theta_best)
array([[0.99530983],
       [0.98646971]])
```

The linear regression now correctly fits the transformed data set, identifying the following model: $\hat{y}' = 0.98 + 0.99x$. Note that the presence of noise renders it impossible to retrieve the exact values of the intercept and the angular coefficient.

Alternatively, a simpler (and more concise) way to perform a linear regression fit is to directly compute the pseudoinverse of the independent variable matrix x (see equation [2.10]). This can be done, in few lines of code, using Scikit–Learn's library (Buitinck et al. 2013):

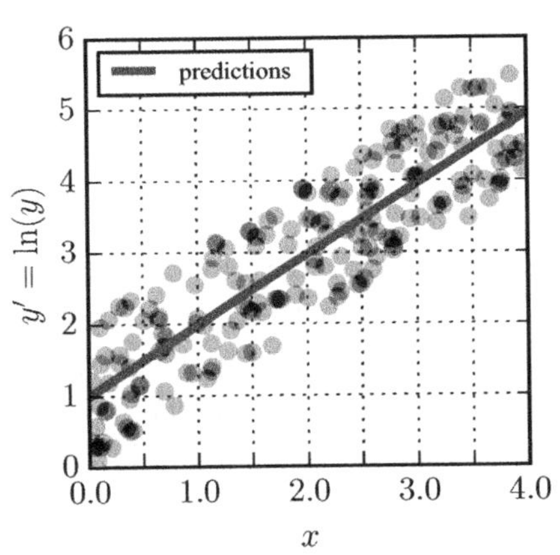

Figure 2.3. *Comparison of the linear regression model predictions with the linearized data set (see equation [2.12])*

```
from sklearn import linear_model
from sklearn.metrics import mean_squared_error
regr = linear_model.LinearRegression()
                              # create linear regression object
regr.fit(X, y_prime) # train the model
y_predict = regr.predict(X) # make predictions
print("Parameters: ", regr.intercept_,regr.coef_) # parameters
print("Mean squared error: %.4f" % mean_squared_error(y_prime,
                                             y_predict)) # MSE
Parameters:  [0.99530983] [[0.98646971]]
Mean squared error: 0.3459
```

2.3. Gradient descent

We have seen that linear regression allows us to fit data sets with pre-identified linear models or hypothesis $f_{\boldsymbol{\theta}}$. The advantages are (i) the admission of analytically tractable, best-fit solutions and (ii) the reduced computational complexity. However, the major drawback is the impossibility to fit nonlinear functions, much more abundant in nature than linear ones (see Chapter 5).

To this end, the general theory of nonlinear regression[4] considers a nonlinear hypothesis function $f_{\boldsymbol{\theta}}(\mathbf{x})$, in contrast with the linear function in equation [2.3]. In this case, if we proceed as we have done for linear regression – that is, requiring that $\partial \text{MSE}(\mathbf{X}, f_{\boldsymbol{\theta}})/\partial \boldsymbol{\theta} = 0$ – we obtain the following nonlinear system of equations:

$$\frac{1}{n}\sum_{k=1}^{n}\left(f_{\boldsymbol{\theta}}(\mathbf{x}^{(k)}) - y^{(k)}\right)\frac{\partial f_{\boldsymbol{\theta}}}{\partial \boldsymbol{\theta}} = 0. \qquad [2.13]$$

Unfortunately, there are no analytical methods to solve such a nonlinear system for a general (unspecified) nonlinear hypothesis $f_{\boldsymbol{\theta}}$. And, actually, sometimes equation [2.13] may even not admit a solution or admit an infinity of solutions. In such scenarios, the idea to solve the nonlinear system [2.13] is to resort to iterative approaches

4 Note, however, that if we formulate the nonlinear regression problem differently, within a reduced special setting, a seemingly linear hypothesis function can actually be formulated (see section 2.5).

which, depending on the good (or bad) initial guess, may converge to the global (or a local) minimum error.

One of the most effective approaches to identify the roots of a nonlinear system of equations is *GD* (Boyd and Vandenberghe 2004). GD, also called steepest descent, is an optimization algorithm for finding a local minimum of a differentiable function – that is, a set of optimal parameters $\boldsymbol{\theta}$ that minimizes a given (differentiable) loss function.

A quite easy way to grasp the essence of GD is to imagine ourself lost in a foggy mountain, where visibility is so limited that we can only sense the slope beneath our feet. To find the way out quickly, a smart approach would be to head downhill, following the steepest slope.[5] This is exactly what GD does: it measures the local gradient of the loss function with regard to the parameters $\boldsymbol{\theta}$ and tweaks the latter to go in the direction of descending gradient. Once the gradient is zero, we have reached the minimum.

Let us break it down further. At the beginning, $\boldsymbol{\theta}$ is filled with random values or some other predefined values. This process is called *initialization*. Then, in small incremental steps, often called *epochs*, the algorithm tweaks repeatedly the parameters to minimize the value of the loss function $E(\boldsymbol{\theta}, f_{\boldsymbol{\theta}})$ (that must be differentiable for all $\boldsymbol{\theta}$ – which also implies that $f_{\boldsymbol{\theta}}$ must be differentiable for every $\boldsymbol{\theta}$) according to the regular GD optimizer equation:

$$\boldsymbol{\theta}^{\text{next step}} = \boldsymbol{\theta} - \eta \frac{\partial E}{\partial \boldsymbol{\theta}} (\boldsymbol{\theta}, f_{\boldsymbol{\theta}}), \qquad [2.14]$$

where η is the *learning rate*. In other words, we need to compute how much the loss function E will change if we change θ_i just a little bit, with $i = 1, 2, \cdots, m$.

By resorting to the analogy of the foggy mountain, it is like asking "What is the slope of the mountain under our feet if we face east?"

5 Disclaimer: if you are truly lost on a mountain, then you should probably stop and wait for a break in the fog, rather than following the steepest descent.

and then asking the same question facing north (and so on for all other dimensions i up to m, if we can imagine a universe with more than three dimensions).

After having computed the gradient vector $\partial E/\partial \boldsymbol{\theta}$ which points uphill, we just have to head in the opposite direction to go downhill – that is, subtracting the gradient from the parameters $\boldsymbol{\theta}$, with a weighting factor η. This process continues until the algorithm converges, reaching a (global or local) minimum value, as shown in Figure 2.4.

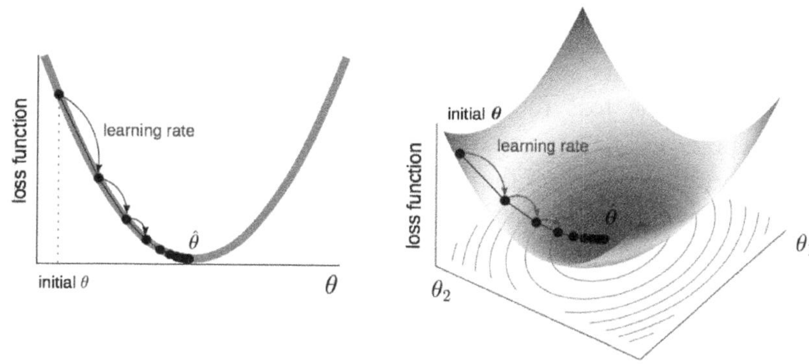

Figure 2.4. *Schematic representation of gradient descent; the model parameters are initialized randomly and get tweaked repeatedly to minimize the cost function; the learning step size is proportional to the slope of the cost function, so the steps gradually get smaller as the parameters approach the minimum*

When optimizing an objective function using GD, two important aspects must be considered:

– The size of the incremental steps, which is determined by the **learning rate**, η. The latter is a *hyperparameter*. A hyperparameter is a variable controlling the learning process (in contrast to the model parameters that are determined via training). The learning rate, as such, is crucial in identifying the best value of the parameters $\boldsymbol{\theta}$ within the smallest number of iterations. Indeed, if η is set too small, the algorithm will require numerous iterations to converge, resulting in a time-consuming computations. Conversely, if η is excessively high, we might overshoot the minimum and end up on the opposite side, potentially at a point where error is higher than before. This scenario could even cause the algorithm to diverge, producing increasingly larger error values and failing to find an optimal solution.

– The **loss function**: Not all loss functions exhibit a smooth, bowl-like shape as the ones depicted in Figure 2.4. Depending on the choice of loss function and the hypothesis function, we may find irregularities such as holes, ridges, plateaus and many local minima, creating challenges for convergence toward the (global) minimum. Also remember that the loss function must always be differentiable,[6] otherwise we cannot compute the optimizer equation [2.14].

Figure 2.5 highlights some of the primary obstacles that we may encounter with some loss functions and, in most of the cases, for nonlinear hypotheses f_{θ} – but also in neural networks (see Chapter 5). When the algorithm starts on the left due to random initialization, it attempts to converge to a local minimum, which may not be as desirable as the global minimum. This same local minimum is a point of discontinuity that renders even more difficult the convergence. On the other hand, if the algorithm begins on the right, it finds a plateau and crossing it takes an extensive amount of time and if we prematurely stop the algorithm we will not succeed in reaching the global minimum.

Nonetheless, the MSE for linear regression models happens to be always differentiable and convex. This means that any two points on the curve can be connected by a line segment that never intersects the curve. Consequently, there are no local minima; we have only one single global minimum. Moreover, the MSE function is continuous, exhibiting a slope that changes gradually (differently from the MAE). These characteristics have a significant implication: given sufficient time and an appropriate value of the learning rate, GD is guaranteed to approach the global minimum closely.

6 In reality, when dealing with a loss function that is not everywhere differentiable, an ad hoc subgradient procedure can be used. Instead of calculating the gradient, we use a subgradient that provides a valid lower bound on the slope of the loss function at that particular singular point.

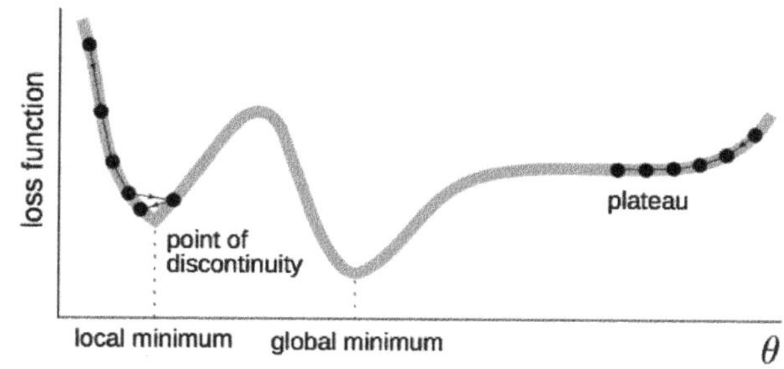

Figure 2.5. *Gradient descent for a general non-convex loss function (readapted from Géron (2022))*

In linear regression, the shape of the MSE loss function always resembles a bowl. However, we should note that the latter can be elongated if the parameters – that is, the features – have different scales. Figure 2.4 (right) demonstrates GD applied to a set with features θ_1 and θ_2, where the latter has significantly smaller values than the former. Note that, in general, when features have quite different scales, the convergence to the global minimum requires a large number of iterations. Such a drawback can be easily overcome by appropriately scaling the features – that is, by scaling the independent and dependent variables (see section 2.4).

It is worth noting that training a model, whether linear or not, involves searching for a perfect combination of model parameters that minimizes the loss function across the entire data set. This search takes place within the vast parameter space of the model. Imagine we are trying to train a model with an extensive set of parameters that surpasses the simplicity of the one- and two-parameters models in Figure 2.4. In this scenario, we are essentially following the same steps as before, but now in a higher dimensional space. The dimension of this space is equal to the number of parameters, making the search much more challenging. Indeed, finding a needle in a haystack that has 1,000 dimensions is significantly more intricate than doing so in just two dimensions. However, in the case of linear regression (unlike artificial neural networks, see Chapter 5), the loss function has optimal properties and the needle (global minimum) always resides comfortably at the bottom.

2.3.1. *Batch GD*

Batch GD is the most cumbersome implementation of the GD algorithm. It involves the use of equation [2.14] for the full training set $\mathbf{X}$ at every step – that is, the whole batch of training data (and this is the reason of its name).

Applying batch GD for a linear regression problem starts with the computation of the partial derivative of the MSE function with respect to the model parameter:

$$\frac{\partial}{\partial \boldsymbol{\theta}} \mathrm{MSE}\left(\mathbf{X}, f_{\boldsymbol{\theta}}\right) = \frac{2}{n} \mathbf{X}^{\mathrm{T}} \left(\mathbf{X}\boldsymbol{\theta} - \mathrm{y}\right), \qquad [2.15]$$

and continues by recursively using equation [2.14] to update the parameters. The recursion stops when we have reached the prescribed number of steps (epochs) or the error has become smaller than a prescribed threshold.

We can now use the training data set from the previous example and find a linear regression model by means of batch GD:

```
eta = 0.05 # learning rate
n_epochs = 500 # no epochs
theta = np.array([[0.],[0.]]) # initialization
for epoch in range(n_epochs):
    gradients = 2/n_snapshots * X_p.T.dot(X_p.dot(theta)
                                - y_prime) # [2.15]
    theta = theta - eta * gradients # optimizer [2.14]
print(theta)
[[0.99530187]
 [0.98647279]]
```

It should come as no surprise that we obtain the same result as using the normal equation. However, we might wonder what would happen if we had chosen a different learning rate. In Figure 2.6, we visualize the predictions of the same linear model within the first 10 iterations using different learning rates, all starting from the same initial point (represented by the dotted line). Similarly, Figure 2.7 illustrates the evolution of the model parameters, θ_1 and θ_2, with the contours representing the loss function (MSE) that has a global minimum approximately located at $(1, 1)$.

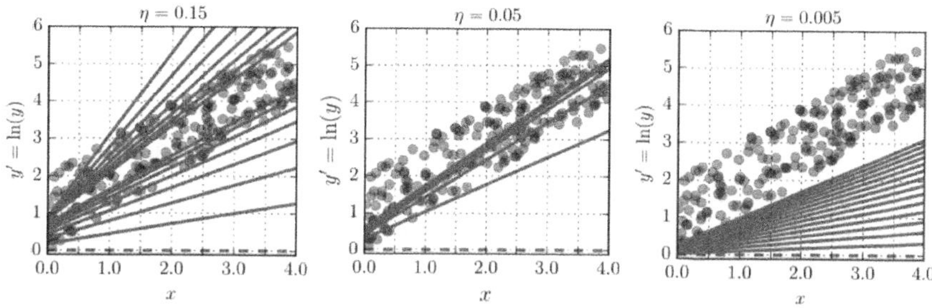

Figure 2.6. *Batch gradient descent with various learning rates with the same initialization parameters*

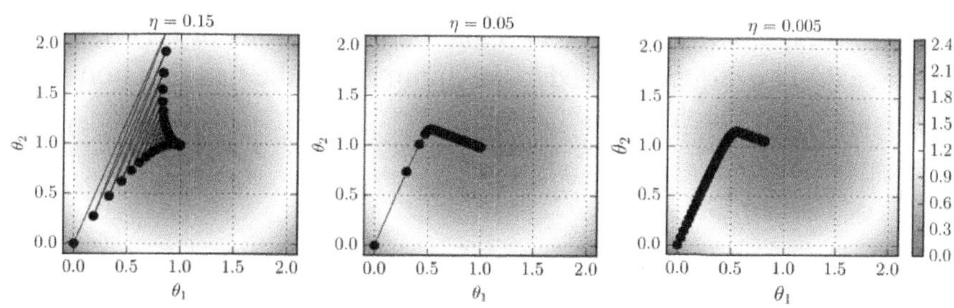

Figure 2.7. *Evolution of the model's parameters during batch gradient descent with various learning rates (see Figure 2.6)*

Starting from the left, we observe that with a relatively high learning rate, the algorithm bounces around during each iteration. In this particular scenario, if we select a larger value of the learning rate, the algorithm will diverge and we will never reach the minimum (you can try by yourself with the above code). In the middle, a moderately appropriate learning rate allows GD to smoothly reach the global minimum. On the right, with a very low learning rate, the algorithm will eventually converge to the solution, but it will take a considerable amount of time.

To identify an appropriate learning rate value, we can employ two strategies: first, we can use an adaptive learning rate (see section 2.3.2) or second, we can fine-tune the learning rate using grid search[7] (Géron 2022).

Despite being much faster than using the normal equation [2.9] or computing the pseudoinverse [2.10] for high-dimensional linear regression tasks, batch GD is the most computationally demanding among all GD variants and the most vulnerable to "bad" initialization. In addition, the evaluation of the optimizer equation [2.14] is often computationally intractable, especially for deep neural networks (see Chapter 5). This is primarily due to two reasons: (i) the parameter vector, collecting the parameters θ_m, can be quite large, and (ii) the number of data points n can also be large. Thus, utilizing the entire training set for computing the gradient can hinder convergence speed.

As it follows, we will explore two alternative versions of GD that can enhance the convergence speed to the minimum of the loss function: stochastic and mini-batch GD.

2.3.2. *Stochastic GD*

Stochastic GD is a version of GD that differs from batch GD in how it estimates the gradient in equation [2.14]. Instead of using all n training data points, stochastic GD randomly selects a single datum (snapshot) from the training set to estimate the gradient at each iteration. As a result, stochastic GD is faster than batch GD: it only needs to process a single instance at a time allowing efficient memory allocation when dealing with large data sets (see also Chapter 5).

However, due to its stochastic nature, stochastic GD exhibits less smooth behavior compared to batch GD. Instead of smoothly

7 Grid search is a technique used in ML to systematically search for the optimal combination of hyperparameter values for a given model. It involves specifying a set of possible values for each hyperparameter and exhaustively evaluating the model performance using all possible combinations of these values. By evaluating the model accuracy for each combination, grid search helps identify the hyperparameter configuration that yields the best performance.

decreasing toward the minimum, the loss function experiences fluctuations, with only an average error decrease over iterations. Consequently, when the algorithm stops, the final parameter values are generally good but not necessarily optimal. Yet, it is worth noting that the inherent randomness in stochastic GD can actually aid in escaping eventual local minima, increasing the chances of finding the global minimum compared to batch GD.

Hereinafter, we implement the stochastic GD for the same above example:

```
eta = 0.05 #learning rate
n_epochs = 20 # no epochs
theta = np.array([[0],[0.]]) #initialization
for epoch in range(n_epochs):
    for i in range(n_snapshots): # iterate over each snapshot
        random_index = np.random.randint(n_snapshots)
                                            # random snapshot
        xi = X_p[random_index:random_index+1]
        yi = y_prime[random_index:random_index+1]
        gradients = 2 * xi.T.dot(xi.dot(theta) - yi)
        theta = theta - eta * gradients # [2.15]
print(theta)
[[0.83905266]
 [1.15041967]]
```

The results are shown in Figure 2.8, where we can observe that the algorithm struggles in converging to the global minimum, rather bouncing all around. This is to the significant dependence of stochastic GD on the learning rate, which in turn is due to inherent stochastic nature of the former.

The problem can be alleviated by gradually adapting the learning rate while training.

Below, we repeat the learning process using a rather simple learning schedule that linearly decreases the learning rate at each epoch:

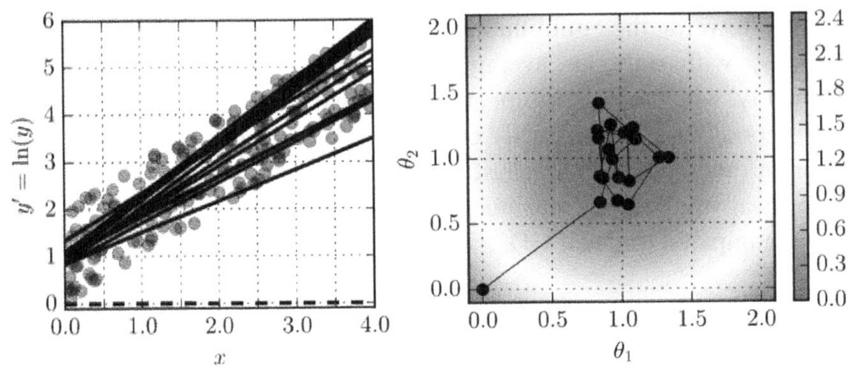

Figure 2.8. *Stochastic gradient descent and evolution of the parameters, with constant learning rate*

```
theta = np.array([[0],[0.]]) #initialization
t0, t1 = 1, 100 # learning schedule hyperparameters
def learning_schedule(t):
    return t0 / (t + t1)
for epoch in range(n_epochs):
    for i in range(n_snapshots):
        random_index = np.random.randint(n_snapshots)
        xi = X_p[random_index:random_index+1]
        yi = y_prime[random_index:random_index+1]
        eta = learning_schedule(epoch * n_snapshots + i)
            # schedule
        gradients = 2 * xi.T.dot(xi.dot(theta) - yi)
        theta = theta - eta * gradients
print(theta)
[[0.90607535]
 [1.02276101]]
```

In 20 iterations only, the algorithm converges – relatively close – to the global minimum, as depicted in Figure 2.9. And this should be compared with the 500 iterations that were needed for the batch GD to reach the minimum.

2.3.3. *Mini-batch GD*

Once we grasped both batch and stochastic GD, understanding mini-batch GD becomes straightforward if we get inspiration from

Aristotle: "*In medio stat virtus*" – that is, "the best option lies between two extremes".

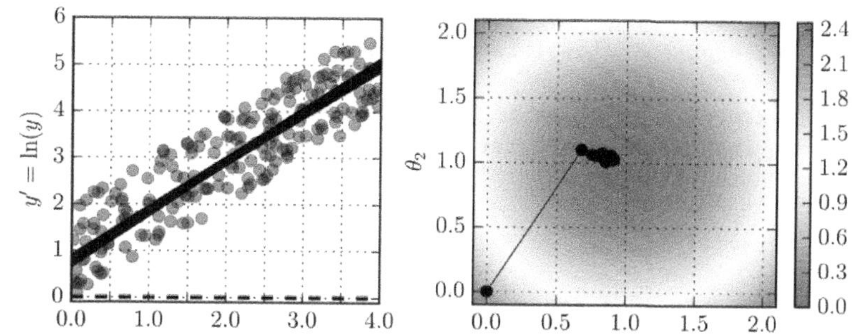

Figure 2.9. *Stochastic gradient descent with learning schedule*

Instead of computing gradients based on the entire training set (batch GD) or a single datum (stochastic GD), mini-batch GD computes gradients on small random sets of data points known as *mini-batches*. This approach offers the advantage of leveraging optimized matrix operations, resulting in improved performance compared to stochastic GD. At the same time, it avoids the need for large memory allocation required by batch GD. Depending on the size of the mini-batch, the optimization path followed by mini-batch GD is less erratic than its stochastic counterpart, often leading to a more optimal minimum.

Implementing mini-batch GD is relatively simple. We only need to make slight modifications to the stochastic GD algorithm and provide an iterator to extract mini-batches:

```
def iterate_minibatches(inputs, targets, batchsize,
          shuffle=False):
    '''Iterator over mini-batches
    :param inputs, targets: independent and dependent variables
    :param batchsize: mini-batch size
    :param shuffle: shuffle mini-batches
    :return: mini-batches of inputs and targets'''
    # Check if no of samples in inputs and targets are equal
    assert inputs.shape[0] == targets.shape[0]
    if shuffle:
        indices = np.arange(inputs.shape[0])  # array of indices
        np.random.shuffle(indices)  # Shuffle indices
```

```
    # Iterate over mini-batches
    for start_idx in range(0, inputs.shape[0] - batchsize+1,
                                                  batchsize):
        if shuffle:  # subset of shuffled indices
            excerpt = indices[start_idx:start_idx + batchsize]
        else: # slice to select samples
            excerpt = slice(start_idx, start_idx + batchsize)
        yield inputs[excerpt], targets[excerpt]
                                              # return mini-batches
eta = 0.05 #learning rate
theta = np.array([[0],[0.]]) #initialization
batch_size = 30 # define mini-batches size
for epoch in range(n_epochs):
    for batch in iterate_minibatches(X_p, y_prime, batch_size,
                                     shuffle=True):
        x_batch, y_batch = batch
        gradients = 2/batch_size *
        x_batch.T.dot(x_batch.dot(theta) - y_batch)
        theta = theta - eta * gradients
print(theta)
[[0.94862053]
 [0.96679924]]
```

It is done much faster and more accurately than the stochastic GD! Figure 2.10 depicts the predictions of the model within the first 10 iterations of the algorithm and the parameters evolution. Note that the value of the minimum reached by the mini-batch GD intrinsically depends on the selected size of the mini-batch. The latter represents yet another important hyperparameter of the optimization problem.

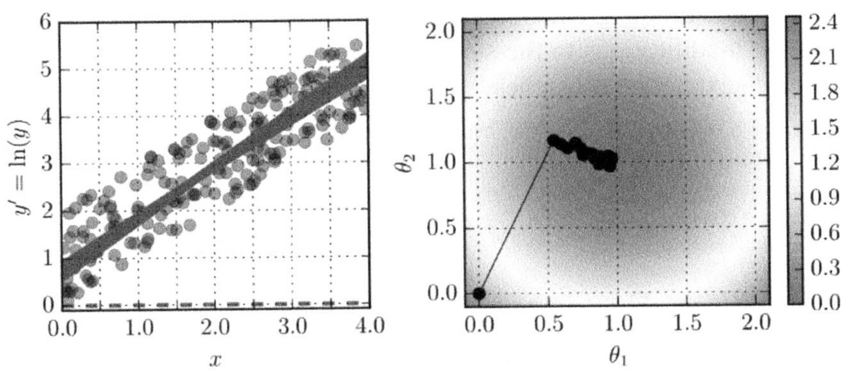

Figure 2.10. *Mini-batch gradient descent*

To perform the same task, we could also use Scikit-Learn, namely, the `SGDRegressor` class, which by default minimizes the MSE. The following code runs for maximum 20 epochs or until the loss drops by less than 0.001 during one epoch (`max_iter=20, tol=1e-3`). It starts with a learning rate equal to `eta` and use the following learning schedule: $\eta^{(\text{next step})} = \eta t^{-1/4}$. The option `penalty=None` avoids the use of regularization strategies (see section 2.6).

```
from sklearn.linear_model import SGDRegressor
sgd_reg = SGDRegressor(max_iter=20, tol=1e-3, penalty=None,
                                            eta0=eta)
for epoch in range(n_epochs):
    for batch in iterate_minibatches(X_p, y_prime, batch_size,
                                        shuffle=True):
        x_batch, y_batch = batch
        sgd_reg.partial_fit(X, y_prime.ravel())
                                        # partial_fit mini-batch
print(sgd_reg.intercept_, sgd_reg.coef_)
[1.00172812] [1.00508048]
```

Also note that if we were performing stochastic GD, we would not need to iterate over the mini-batches, and we should replace `partial_fit()` with `fit()`.

In conclusion, the three alternative algorithms for GD – batch, stochastic and mini-batch – are powerful tools for solving an optimization problem and they should be preferred to the normal equation and the computation of the pseudoinverse, when dealing with a high-dimensional linear regression problem (with large numbers of data sets or large number of parameters).

The main difference among the three alternatives is summarized by Figure 2.11, which provides a comparison for the parameters trajectory during training. The mini-batch GD is often preferred as it allows a good compromise in terms of the goodness of the reached minimum and the computational speed; however, it is difficult to provide strict recommendations for a general case. As a general rule, we may opt for a mini-batch GD and train the model with different mini-batch sizes – if equal to 1, we will resort to stochastic GD, and if equal to the training set size, we will end up with batch GD.

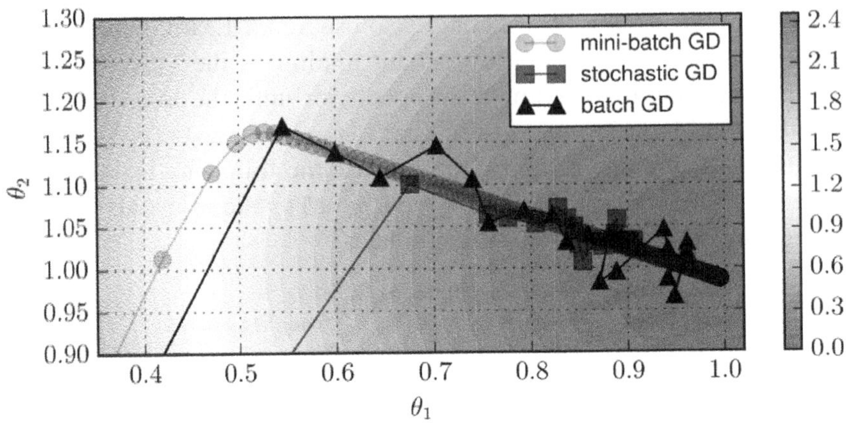

Figure 2.11. *Gradient descent paths comparison in the parameters space*

Finally, it is worth noticing that herein we limited ourselves to the standard optimizer equation [2.14]; however, alternative and more effective solutions exist (see Chapter 5).

2.4. Data preprocessing and model validation

In the above examples, we have seen how to perform linear regression given some training data sets. However, we have neglected two important aspects common to all ML models – regression included *feature scaling* and models *validation*. The former is part of the preprocessing to train a model, while the latter can be interpreted as postprocessing of the trained model.

2.4.1. *Feature scaling*

Feature scaling is a crucial step in preparing data for ML. Indeed, all GD algorithms and in general most of the optimization algorithms struggle in dealing with variables with different scales (e.g. Figure 2.4). In the frame of GD, there are two common methods to address this issue: normalization and standardization.

Both techniques transform a feature x into its scaled counterpart $\bar{x}$, according to:

$$\bar{x} = \frac{1}{\alpha}\left(x - \beta\right). \qquad [2.16]$$

Normalization shifts and re-scales the feature values to a range between -1 and 1 by considering $\alpha \equiv \frac{1}{2}(\max(x) - \min(x))$ and $\beta \equiv \alpha + \min(x)$. On the other hand, standardization transforms the feature in a distribution with a zero mean and unit variance, by considering $\alpha \equiv \sigma(x)$ and $\beta \equiv \mu(x)$, with μ and σ being the mean and standard deviation, respectively.

Normalization is simple and confines values to a specific range, while standardization is not bounded to a range and is less affected by outliers. Whether it is more appropriate to normalize or *standardize* the data depends mainly on the loss function and the statistics of the data themselves – when the latter are uniformly distributed, normalization is to be preferred; in all the other cases, standardization is the best choice.

Hereinafter an explicit implementation of both scaling techniques, depending on how the function is called:

```
def scaling(x, fit=False, transform=False,
                    inverse_transform=False,
            norm=True, param=None):
    ''' scale variable x
        :param x: variable
        :param fit: find scaling parameters
        :param transform: scale x
        :param inverse_transform: inverse scale x
        :param norm: inverse scale x
        :param fit: scaling parameters
        :return: scaling parameters (if fit=False)
                 scaled x (if transform=True)
                 inverse scaled x (if inverse_transform=True) '''
    if fit==True:
        if norm==True:
            min_ = np.amin(x); max_ = np.amax(x)
            a = 0.5*(max_-min_)
            b = 0.5*(max_+min_)
```

```
        else:
            a = np.std(x)
            b = np.mean(x)
        return [a,b]
    elif transform==True:
        return np.divide(x-param[1],param[0])
    elif inverse_transform==True:
        return np.multiply(x,param[0])+param[1]
```

One could also use directly Scikit-Learn's `preprocessing` module.

2.4.2. *Test and validation of a model*

In the examples above, we discovered regression models that best fit some training data $\{\mathbf{X}, \mathbf{y}\}$. However, in the general case, and not only in the frame of regression methods, we need to evaluate how well a (regression) model performs on new data. To this end, we must first define three notions: *interpolation*, *generalization* and *extrapolation*.

DEFINITION 2.1.– Interpolation is the ability of predicting values within the range of observed data points.

In the context of regression, it means predicting the value of the dependent variables for values of independent variables that lie within the range of the training data.

DEFINITION 2.2.– Generalization is the ability of making predictions for new, previously unseen data, drawn from the same distribution as the ones used to train the model.

ML is all about having models that can generalize well. Trained to solve one problem, the model attempts to utilize the patterns/relationships learned from that task to solve the same task, with slight variations. While the existence of a subtle difference between interpolation and generalization, both terms are often adopted interchangeably.

DEFINITION 2.3.– Extrapolation is the ability of predicting values beyond the range of the observed, training data points.

Extrapolation requires a solid understanding of the relationships identified by a model to apply the same outside the familiar data range.

It is challenging for current ML models to reliably extrapolate, as they often specialize in specific tasks and struggle with broader applications. Many artificial intelligence methods are inherently interpolative, and constructing extrapolative or "intelligent" algorithms still remains an open challenge (see section 2.7 and Chapter 5 of volume 1 and Chapter 3 of volume 2).

From the aforementioned definitions, it is clear that we should *always test models* in order to quantify their generalization capabilities. A cumbersome approach consists of dividing the original data set into two: a *training set* and a *test set*.

The training set is used to train the model, as we have already seen, while the test set is used to assess its performance. The latter is estimated by means of the generalization error, which measures the error of the trained model on the (unseen) test data set. Thus, if a model has low training error but high generalization error, it indicates *overfitting* to the training data. Overfitting is defined as the phenomenon where a model becomes overly complex and excessively fits the training data set, capturing noise and irrelevant patterns instead of learning the true underlying patterns and relationships.

However, evaluating the generalization error multiple times on the same test set and optimizing the model and hyperparameters (such as the learning rate in GD, or the mini-batch size in mini-batch GD) specifically for that set may be suboptimal. Indeed, the model will unlikely perform as well on new data.

2.4.2.1. *Hold-out validation*

To mitigate the problem arising from testing a model on the same test set, a common solution is *hold-out validation*. A portion of the training set is reserved as a *validation set*. Multiple candidate models with different hyperparameters are trained on the reduced training set, and the one that performs best on the validation set is selected. The best model is then trained on the full training set, including the validation set, to create the final model. The performance of the latter is then evaluated on the test set to quantify the generalization error.

While hold-out validation is effective, the size of the validation set is crucial. If it is too small, evaluations may be imprecise, leading to the selection of suboptimal models. Conversely, if the validation set is too large, the remaining training set becomes significantly smaller, making it unfair to compare models trained on a much smaller set.

As a general rule (and a first good guess), we often split data 60%–20%–20% for training–validating–testing. We can easily do it by leveraging Scikit-Learn:

```
from sklearn.model_selection import train_test_split
# split into training+validation and test sets
X_tv, X_tv, y_train, y_test = train_test_split(X, y,
                                        test_size=0.2)
# split into training and validation sets - note: 0.25 * 0.8 = 0.2
X_train, X_val, y_train, y_val =
                        train_test_split(X_tv, y_tv, test_size=0.25)
```

2.4.2.2. *Cross-validation*

Repeated hold-out validation addresses the issue related to the choice of the validation set size by using multiple small validation sets. This process is called k-fold cross-validation, where the training data set is further partitioned into k-folds, which are typically randomly selected portions of the original set. In k-fold cross-validation, the training data are randomly partitioned into k partitions (or folds). Each partition is used to construct a regression model $\hat{y}_{(l)} = f_{\boldsymbol{\theta}_{(l)}}(\mathbf{x}_{(l)})$ for $l = 1, 2, \cdots, k$. Then, the final, cross-validated model is constructed by averaging the values of the parameters obtained from each fold – that is, $\boldsymbol{\theta} = 1/k \sum_{l=1}^{k} \boldsymbol{\theta}_{(l)}$.

Despite its great advantage over hold-out validation, cross-validation may increase tremendously the learning process as each model needs to be trained on different validation sets. We will implement cross-validation both using Scikit-Learn (see example in section 2.5.1) and performing the aforementioned steps explicitly (see application in section 2.7.1).

2.5. Nonlinear regression

Up to this point, we have seen how to train and validate linear models that best fit some data. However, what if the data are more complex and cannot be fitted, with good approximation, by a simple straight line?

In that case, we may simply resort (spoiler alert!) to artificial neural networks (see Chapter 5). But, no worries! A much simpler alternative exists: we can actually fit a linear model to nonlinear data. To this end, we just need to replace the original set of independent variables, $\mathbf{x}$, with a library $\phi(\mathbf{x}) = \left[\phi_1(\mathbf{x}), \phi_2(\mathbf{x}), \cdots, \phi_l(\mathbf{x})\right]^{\mathrm{T}}$ of l candidate basis functions, for example, polynomials, trigonometric or radial basis functions, or any other user-supplied function. In so doing, equation [2.4] becomes:

$$\hat{y} = \boldsymbol{\theta}^{\mathrm{T}} \phi(\mathbf{x}). \qquad [2.17]$$

Everything is settled and we can move straightforward to an example. This time, we will also take care of preprocessing the data (see section 2.4). We will consider as candidate basis functions only polynomial ones: the regression is thus said to be *polynomial*.

2.5.1. *End-to-end example*

Let us consider a data set given by:

$$y = \sin(|x|) + \sin(x^2) + \mathcal{N}(0, \sigma), \qquad [2.18]$$

where $\mathcal{N}(0, \sigma)$ is a normally distributed random variable with mean zero and standard deviation σ. To generate the data, we use the code as follows:

```
import matplotlib.pyplot as plt
import numpy as np
np.random.seed(42)
n_snapshots = 200
noise = np.random.normal(0,0.5,(n_snapshots,1))
                              # generate normal noise
X = np.random.uniform(-np.pi,np.pi,(n_snapshots,1))
y = np.sin(np.abs(X))+np.sin(X)**2+ noise
```

Now, we proceed by splitting the data set into a training and test set using Scikit-Learn's `train_test_split`. Then, we scale both independent and dependent variables based on the statistics of the training data set (see section 2.4).

```
from sklearn.model_selection import train_test_split
# split into training and test sets
X_train, X_test, y_train, y_test =
                    train_test_split(X, y, test_size=0.2)
# compute scaling params
param_x = scaling(X_train,fit=True) # scaling params for X
param_y = scaling(y_train,fit=True) # scaling params for y
# scale X (training and test sets)
norm_X_train = scaling(X_train,transform=True,param=param_x)
norm_X_test = scaling(X_test,transform=True,param=param_x)
# scale y (training and test sets)
norm_y_train = scaling(y_train,transform=True,param=param_y)
norm_y_test = scaling(y_test,transform=True,param=param_y)
```

At this point, we have to construct the polynomial basis functions, with Scikit-Learn's `PolynomialFeatures` (we stop at the sixth degree). Then, we can simply perform a linear regression:

```
from sklearn.preprocessing import PolynomialFeatures
from sklearn.linear_model import LinearRegression
poly_features = PolynomialFeatures(degree=6, include_bias=False)
norm_X_poly = poly_features.fit_transform(norm_X_train)
lin_reg = LinearRegression()
lin_reg.fit(norm_X_poly,norm_y_train) # perform linear regression
lin_reg.intercept_, lin_reg.coef_ # theta_0 and theta_m
(array([-0.35987146]),
 array([[ 0.20041976,   7.23777176,  -0.72575099, -16.10685101,
          0.48763306,   8.95138004]]))
```

It is not so difficult. The model finds the following best-fit equation:

$$\hat{y} = -0.36 + 0.20x + 7.24x^2 - 0.73x^3 - 16.11x^4$$
$$+0.49x^5 + 8.95x^6.$$

We can now deploy it to make predictions and compare the latter with the test set:

```
x = np.expand_dims(np.linspace(-np.pi,np.pi,200),1)
norm_x = scaling(x,transform=True,param=param_x)
norm_x_poly = poly_features.fit_transform(norm_x)
norm_y_predicted = lin_reg.predict(norm_x_poly)
```

Figure 2.12 (left) compares the predictions with the training and test sets and below we evaluate the MSE of the predictions for the training and test sets:

```
from sklearn.metrics import mean_squared_error
norm_X_test_poly = poly_features.fit_transform(norm_X_test)
norm_y_predicted_train = lin_reg.predict(norm_X_poly)
norm_y_predicted_test = lin_reg.predict(norm_X_test_poly)
print('MSE on train set: ',
      mean_squared_error(norm_y_predicted_train, norm_y_train))
print('MSE on test set: ',
      mean_squared_error(norm_y_predicted_test, norm_y_test))
MSE on training data set:  0.05173368072737507
MSE on test data set:  0.05142192371233674
```

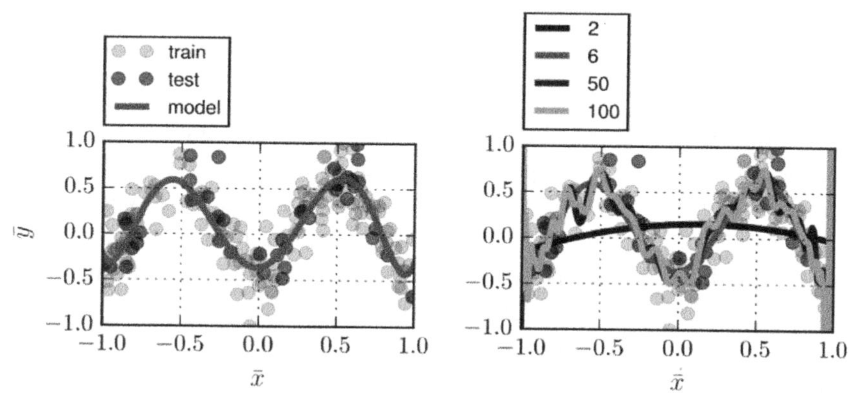

Figure 2.12. *Nonlinear regression using polynomial basis functions with degree six (left) and different degrees (right). Features are scaled*

The model performs quite well on both sets. However, we skipped two important aspects: (i) hyperparameters selection and (ii) validation of the model.

2.5.1.1. *Hyperparameters selection*

The degree of the polynomial basis is actually a hyperparameter. Thus, we cannot a priori say that the analysis is completed – that is, whether or not we found the best model for the data set at hand. To this end, we need to train multiple models with different degrees:

```
degree = [2,6,50,100] # different polynomial degrees
color = ['black','red','darkblue','darkorange']
fig = plt.figure(figsize=(3., 2.))
plt.plot(norm_X_train,norm_y_train,'ko',alpha=0.2)
plt.plot(norm_X_test,norm_y_test,'ko')
for i in range(len(degree)):
    poly_features = PolynomialFeatures(degree=degree[i],
                                       include_bias=False)
    norm_X_poly = poly_features.fit_transform(norm_X_train)
    lin_reg = LinearRegression() # linear regression model
    lin_reg.fit(norm_X_poly,norm_y_train) # fit model
    norm_x_poly = poly_features.fit_transform(norm_x)
    norm_y_predicted = lin_reg.predict(norm_x_poly)
    plt.plot(norm_x,norm_y_predicted,'-',color=color[i],
             linewidth=3,label=str(degree[i]))
plt.ylim(-1,1)
plt.xlim(-1,1)
plt.show()
```

The results are depicted in Figure 2.12 (right) and from them we can draw some conclusions: (i) a quadratic model is unable to fit the data and (ii) high degree – 50 and 100, here – polynomial models wiggle around to get as close as possible to the training data points. In other words, the quadratic model is underfitting, while for high polynomial degrees, the model tends to overfit the training set. So, how can we select the best models – that is, the one that will generalize best (see section 2.4)? To answer let us move to the second part of this example: model validation.

2.5.1.2. *Validation*

We have already seen in section 2.4 that a crucial aspect in finding models that generalize well is to perform hold-out validation. To this end, we should assess the model's performance by examining the validation error at varying of the size of the training set (and eventual hyperparameters). By repeatedly training the model on subsets of varying sizes from the training set, we can generate informative plots that will guide us in selecting the model with the minimum error on the validation set. Below we implement a function that is designed to plot the training, validation and test errors of models with different polynomial degrees (up to 40).

```
def learning_curves(X_train, y_train, X_test, y_test, degree):
    degree +=1
    X_train, X_val, y_train, y_val =
           train_test_split(X_train, y_train, test_size = 0.25)
    norm_X_val = scaling(X_val,transform=True,param=param_x)
                                                        # scale
    model = LinearRegression()
    train_errors = np.zeros((degree-1,len(X_train)-1))
    val_errors = train_errors.copy()
    test_errors = train_errors.copy()
    for deg in range(1,degree):
        for m in range(1, len(X_train)):
            # Construct polynomial library up to degree = deg
            poly_features = PolynomialFeatures(degree=deg,
                                          include_bias=False)
            X_train_poly =
                         poly_features.fit_transform(norm_X_train)
            X_val_poly = poly_features.fit_transform(norm_X_val)
            X_test_poly = poly_features.fit_transform(norm_X_test)
            # Fit model
            model.fit(X_train_poly[:m], y_train[:m])
            # Predict training, validation, and test
            y_train_pred = model.predict(X_train_poly[:m])
            y_val_pred = model.predict(X_val_poly)
            y_test_pred = model.predict(X_test_poly)
            # Store errors
            train_errors[deg-1,m-1] =
                  mean_squared_error(y_train[:m], y_train_pred)
            val_errors[deg-1,m-1] =mean_squared_error(y_val,
                                                      y_val_pred)
            test_errors[deg-1,m-1] = mean_squared_error(y_test,
                                                      y_test_pred)
    return train_errors, val_errors, test_errors
train, val, test = learning_curves(norm_X_train, norm_y_train,
                          norm_X_test, norm_y_test, degree=40)
```

Let us first have a look at the error on the training and validation sets for a sixth-degree polynomial model, depicted in Figure 2.13. We start by analyzing the behavior of the training error. When there is a small bunch of training points, the model can perfectly fit those data and results in a zero error. However, as more points are added, the model struggles to fit the data precisely due to the noise that we artificially added and the intrinsic nonlinearity of the function we are trying to fit.

Thus, the error on the training data increases until it reaches a plateau where additional data will not significantly affect the mean error.

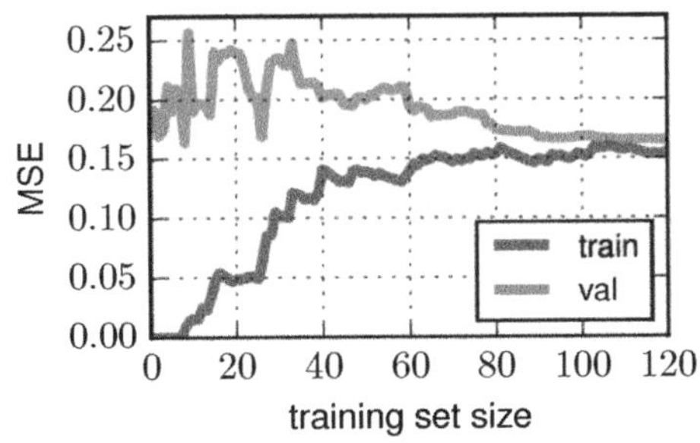

Figure 2.13. *Training and validation MSE at varying training set sizes for a sixth-degree polynomial model*

An analysis on the validation error however rapidly reveals that things are quite different. When there is a small bunch of training points, the model clearly fails to generalize well and leads to relatively high validation errors. As the model is exposed to more training points, it gradually learns and improves, causing the validation error to decrease and, in this example, to reach approximately the same plateau of the training error.

We are now ready to look at the big picture (see Figure 2.14) where we can see the variation of the training, validation and (optionally) test errors at both the polynomial degree and the size of the training set.

The training error presents the same aforementioned behavior. Additionally, we can observe that for any polynomial degree smaller than 6, the error is high – that is, the model is underfitting the training data. Instead, as we increase the degree of the model, we observe a slight reduction of the error (caused by overfitting the training data).

If we move to the validation error, we observe that for training sets smaller than a certain size, the model does not generalize well (exactly as before). However, in addition, we (re-)discover overfitting. For degrees larger than 20, the model displays high validation error meaning that it overfits the training data and cannot accurately predict any other data different from those contained in the training set (poor generalization).

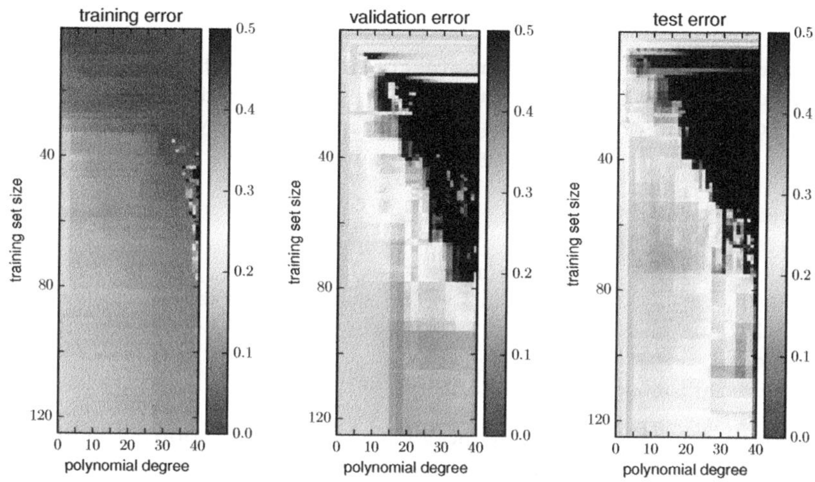

Figure 2.14. *Training (left), validation (center) and test (right) mean squared errors at varying of the training set size and the polynomial degree*

As we have discussed in section 2.4, the validation set allows us to pick up the best model as the one that presents the minimum validation error – approximately corresponding to a degree equal to 16, in this example. Once we have selected such model, we can deploy it to make prediction of new/unseen data – that is, the test set. And, indeed, the minimum error for the test set (i.e. the minimum of the generalization error) is located within the region where the validation error is minimum (see Figure 2.14).

An alternative to the above laborious evaluation of the model performances at varying of the training set size consists of using Scikit-Learn's k-fold cross-validation.[8] The code below, considering a polynomial regression of degree six, splits randomly the training set into five distinct folds, then it trains and evaluates the model five times,

8 Note that `cross_val_score` expects an objective function to be maximized, so the scoring function is actually the opposite of the MSE.

picking a different fold for evaluation every time and training on the other four folds. The result is an array containing the five evaluation scores:

```
from sklearn.model_selection import cross_val_score
poly_features = PolynomialFeatures(degree=6, include_bias=False)
X_train_poly = poly_features.fit_transform(norm_X_train)
scores = cross_val_score(lin_reg, X_train_poly, norm_y_train,
                         scoring="neg_mean_squared_error",
                         cv=5)
mse_scores = -scores
print('Scores: ', mse_scores)
print('Mean scores: ', np.mean(mse_scores))
print('Standard deviation scores: ', np.std(mse_scores))
Scores:  [0.09159855 0.07245614 0.04068157 0.04406373 0.03980408]
Mean scores:  0.05772081499906291
Standard deviation scores:  0.0207975583120194
```

The sixth-degree polynomial model has a MSE of approximately 0.057 ± 0.021 and we would have missed such information if we have just used one validation set.

2.6. Regularization techniques

2.6.1. *Over- and under-determined systems*

In linear and nonlinear regression, we are often confronted with a system of equations, either [2.4] or [2.17], that is under- or over-determined (Gentle 2012). To investigate these scenarios, let us rewrite equation [2.10] as:

$$\mathbf{A}\boldsymbol{\theta} = \mathbf{b}, \tag{2.19}$$

where $\mathbf{A} \equiv \mathbf{X}^{\mathrm{T}}$ is a matrix with n rows and m columns and $\mathbf{b} \equiv \mathbf{y}^{\mathrm{T}}$ is a matrix with n rows and p columns. Note that an analogous formulation can also be written for nonlinear regression [2.13].

With reference to equation [2.19], we define *overdetermined systems* as those systems of equations that have more constraints than unknowns (variables) – that is, when the matrix $\mathbf{A}$ is tall-skinny

– while *underdetermined systems* have more unknowns than constraints – that is, **A** is short-fat (see Figure 2.15).

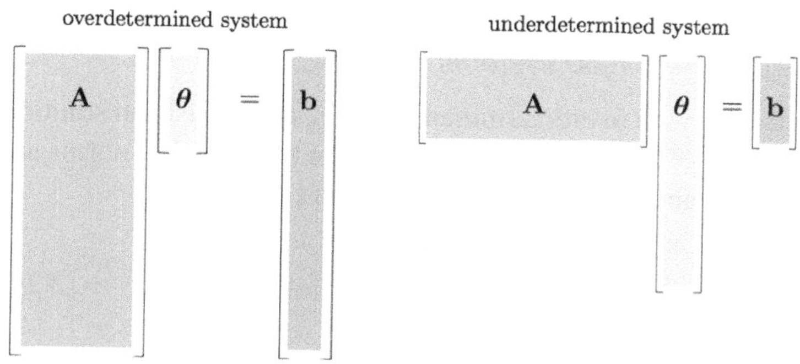

Figure 2.15. *Overdetermined (left) and underdetermined (right) system*

Let us translate these notions from linear algebra to regression methods. If the number of data points is larger than the dimension of the independent variables ($n > m$), then we are dealing with an overdetermined systems of linear equations. Vice versa, if $n < m$, we are confronted with an underdetermined system. In the former case, there are generally no solutions satisfying the linear system, and instead, approximate solutions are found to minimize a given error. In the latter, there is an infinite number of solutions, and some additional constraints must be enforced in order to select an appropriate solution.

Note that in regression, we are most often dealing with overdetermined systems, even if sometimes overparameterization is preferred (Bartlett et al. 2020). Analogous underdetermined systems are instead met when dealing with overparameterized artificial neural networks (see Chapter 5) (Jiang et al. 2019).

Overdetermined or underdetermined optimization problems for linear systems [2.19] involve the minimization of the error of the solution, the MSE for instance, plus a constraint ($n < m$) or a penalty ($n > m$), which is also known as *regularization*, that is:

$$\hat{\boldsymbol{\theta}} = \underset{\boldsymbol{\theta}}{\operatorname{argmin}} \Big(\mathrm{MSE}\,(\mathbf{A}\boldsymbol{\theta} - \mathbf{b}) + \lambda\, w(\boldsymbol{\theta}) \Big), \qquad [2.20]$$

where λ is a weighting parameter and $w(\boldsymbol{\theta})$ a given function, depending whether the system is over- or under-determined. Below, we will see how ℓ_1 and ℓ_2 norm penalties/constraints can help in solving $\mathbf{A}\boldsymbol{\theta} = \mathbf{b}$.

2.6.1.1. *Overdetermined systems*

When dealing with overdetermined systems, there exists no solution that satisfies $\mathbf{A}\boldsymbol{\theta} = \mathbf{b}$. Thus, penalties based on the ℓ_1 and ℓ_2 norms are often considered and the optimization problem [2.20] reads:

$$\hat{\boldsymbol{\theta}} = \underset{\boldsymbol{\theta}}{\operatorname{argmin}} \left(\text{MSE}\left(\mathbf{A}\boldsymbol{\theta} - \mathbf{b}\right) + \lambda_1 ||\boldsymbol{\theta}||_1 + \lambda_2 ||\boldsymbol{\theta}||_2 \right), \qquad [2.21]$$

where $||\boldsymbol{\theta}||_r = \left(|\boldsymbol{\theta}|^r\right)^{1/r}$ denotes the ℓ_r norm and the parameters λ_1 and λ_2 control the penalization of the ℓ_1 and ℓ_2 norms, respectively. The above penalty is crucial for obtain optimal approximate solutions, which are directly affected by the value of the penalty parameters.

To fix the ideas, consider an overdetermined system of equations. We draw $\mathbf{A}$ and $\mathbf{b}$ from normal distributions $\mathcal{N}[0, 1)$. We then obtain the parameters $\boldsymbol{\theta}$ by (i) using equation [2.10] and (ii) solving the optimization problem [2.21], considering two combinations of penalty parameters: $\{\lambda_1, \lambda_2\} = \{0.1, 0\}, \{0, 0.1\}$.

Figure 2.16 shows the obtained $\hat{\boldsymbol{\theta}}$, as a $m \times p$ matrix. The key difference between the two regularization techniques and the ordinary solution without regularization is that ℓ_1 penalty shrinks the parameters associated with less important variables to zero. Indeed, regularization based on the ℓ_1 norm promotes a parsimonious solution, dominated by zero entries, that is, sparse, while regularization based on the ℓ_2 norm keeps the solution values as small as possible. We could also combine both kinds of regularization to promote both sparsity and decrease in the value of the parameters.

2.6.1.2. *Underdetermined systems*

When dealing with undetermined systems, there exists an unlimited set of potential solutions that satisfy the equation $\mathbf{A}\boldsymbol{\theta} = \mathbf{b}$. In such cases, the objective is to introduce constraints that would ideally result in a single, unique solution among the countless possibilities. Usually, the optimization problem [2.20] is reformulated as:

$$\hat{\boldsymbol{\theta}} = \underset{\boldsymbol{\theta}}{\operatorname{argmin}} \left(\text{MSE}\left(\mathbf{A}\boldsymbol{\theta} - \mathbf{b}\right) + ||\boldsymbol{\theta}||_r \right), \qquad [2.22]$$

where the r denotes the r-norm. Once again, one could use either ℓ_1 or ℓ_2 norm, or a combination of them, as for the overdetermined case.

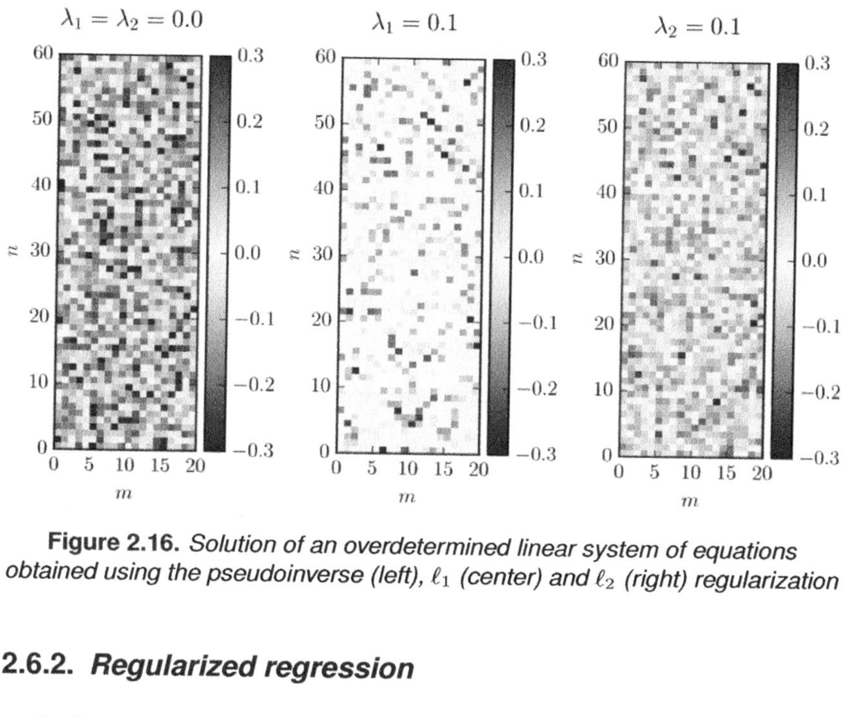

Figure 2.16. *Solution of an overdetermined linear system of equations obtained using the pseudoinverse (left), ℓ_1 (center) and ℓ_2 (right) regularization*

2.6.2. *Regularized regression*

It is now time to see how the aforementioned penalties and constraints for the solution of under- and over-determined systems can be employed in regression.

As already mentioned, (non)linear regression deals, most of the time, with overdetermined systems of equations – all examples we have looked up so far are indeed characterized by $n \gg m$. We should not be surprised, at this point, that the same penalty terms that are used for the solution of overdetermined systems have their own counterparts (and names) in regression. Regularized regression models based on the ℓ_1 and ℓ_2 norms are called Least Absolute Shrinkage and Selection Operator Regression (LASSO) and Ridge, respectively. The combination of both is often referred to as Elastic net.

2.6.2.1. *LASSO*

LASSO regression (Tibshirani 1996) is the (non)linear regression version with a regularization term proportional to $||\boldsymbol{\theta}||_1$. This forces the learning algorithm to not only fit the data but also to shrink the model parameters to zero. The loss function is thus defined as:

$$\mathcal{L}^{\text{LASSO}}\left(\mathbf{X}, f_{\boldsymbol{\theta}}\right) = \text{MSE}\left(\mathbf{X}, f_{\boldsymbol{\theta}}\right) + \lambda_1 ||\boldsymbol{\theta}||_1. \qquad [2.23]$$

Note that the above loss function is not differentiable at $\boldsymbol{\theta} = \mathbf{0}$, thus we cannot derive a closed form solution analogous to the normal equation. However, we can still use GD provided that we define the following modified gradient, whenever $\boldsymbol{\theta}$ is equal to zero:

$$\text{grad}\left(\boldsymbol{\theta}, \mathcal{L}^{\text{LASSO}}\right) = \frac{\partial}{\partial \boldsymbol{\theta}} \text{MSE}(\mathbf{X}, f_{\boldsymbol{\theta}}) + \text{sign}\left(\boldsymbol{\theta}\right), \qquad [2.24]$$

where $\text{sign}(\cdot)$ is the sign function. Below, we perform LASSO regression using Scikit-Learn's for the same example we have seen for polynomial regression (see section 2.5) and we consider polynomials of degree 100:

```
from sklearn.linear_model import Lasso
lambda_1 = 0.0001
lasso_reg = Lasso(alpha=lambda_1,max_iter=100000)
                                        # define lambda_1
# Polynomial feature up to degree 100
poly_features = PolynomialFeatures(degree=100, include_bias=False)
lasso_reg.fit(norm_X_poly,norm_y_train) # fit LASSO reg model
norm_x_poly = poly_features.fit_transform(norm_x)
norm_y_predicted_lasso = lasso_reg.predict(norm_x_poly)
                                        # make predictions
```

The predictions are shown in Figure 2.17 (left). For the sake of clarity, the plot also shows the solution obtained without regularization (pinv). We can easily observe that LASSO allows to rediscover the underlying function (equation [2.18]) used to generate the data, independently of the additive noise and without overfitting – in contrast with the ordinary regression model (pinv), with all polynomial degrees up to 100. The reason is that LASSO shrinks most parameters to zero, hence we discover a quite precise series expansion of the combination of sinusoidal functions within the range $(-\pi, \pi)$.

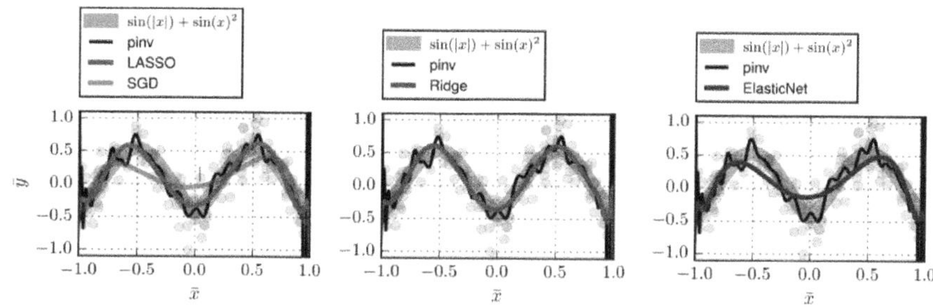

Figure 2.17. *Nonlinear regression using polynomials of degree 100. Comparison between standard regression (pinv) – equation [2.10] – and regularized models: LASSO and stochastic GD (left), Ridge (center) and Elastic net (right). Features are scaled*

For the sake of completeness, note that we could also use Scikit-Learn's `SGDRegressor` with the following arguments: `penalty="l1"` and `alpha=lambda_1`:

```
from sklearn.linear_model import SGDRegressor
sgd_reg = SGDRegressor(max_iter=10000,
                       tol=1e-6,penalty="l1",alpha=lambda_1,
                       eta0=0.01)
sgd_reg.fit(norm_X_poly,norm_y_train.ravel())
norm_y_predicted_sgd = sgd_reg.predict(norm_x_poly)
                                             # make predictions
```

2.6.2.2. *Ridge*

Ridge regression (Hoerl and Kennard 1970) is the (non)linear regression version with a regularization term proportional to $||\boldsymbol{\theta}||_2^2$ that forces the model parameters to be as small as possible. The loss function is defined as:

$$\mathcal{L}^{\text{Ridge}}(\mathbf{X}, f_{\boldsymbol{\theta}}) = \text{MSE}(\mathbf{X}, f_{\boldsymbol{\theta}}) + \lambda_2||\boldsymbol{\theta}||_2^2. \quad [2.25]$$

As with linear regression – and differently from LASSO – we can perform Ridge regression either by computing a closed-form equation (Hoerl and Kennard 1970) or by performing GD. The code hereinafter implements the closed-form solution using Scikit-Learn's `Ridge`:

```
from sklearn.linear_model import Ridge
lambda_2 = lambda_1
ridge_reg = Ridge(alpha=lambda_2,max_iter=100000)
                                          # define lambda_2
ridge_reg.fit(norm_X_poly,norm_y_train) # fit Ridge reg model
norm_y_predicted_ridge = ridge_reg.predict(norm_x_poly)
                                   # make predictions
```

We plot the predictions in Figure 2.17 (center). Once more, thanks to regularization, we are able to retrieve the original function, in contrast with the massively overfitted linear model (pinv).

2.6.2.3. *Elastic net*

Elastic net is the middle version between LASSO and Ridge. Accordingly, the regression model is regularized with respect to both the ℓ_1 and ℓ_2 norms. The regularization term is a mix of both Ridge and LASSO controlled by a ratio ϱ:

$$\mathcal{L}^{\text{Elastic net}}\left(\mathbf{X}, f_{\boldsymbol{\theta}}\right) = \text{MSE}\left(\mathbf{X}, f_{\boldsymbol{\theta}}\right) + \varrho\lambda||\boldsymbol{\theta}||_1 + \frac{1-\varrho}{\varrho}\lambda||\boldsymbol{\theta}||_2^2. \quad [2.26]$$

Elastic net is equivalent to Ridge if $\varrho = 0$ and to LASSO if $\varrho = 1$. Below, we provide the code to define and fit an Elastic net model, whose predictions are shown in Figure 2.17 (right):

```
from sklearn.linear_model import ElasticNet
net_reg = ElasticNet(alpha=2*lambda_2,l1_ratio=1.0,
                     max_iter=100000) #l1_ratio = \varrho
net_reg.fit(norm_X_poly,norm_y_train) # fit Elastic Net reg model
norm_y_predicted_net = net_reg.predict(norm_x_poly)
                                   # make predictions
```

2.7. Challenges in generalization and extrapolation

Regression, like other supervised ML methods, is mainly design for making predictions rather than determining existing relationships between some features. However, by post-processing the model parameters $\boldsymbol{\theta}$, we can, under some circumstances, grasp the underlying relationships buried under the mere data points we are analyzing. In this sense, we may even end up (re)discovering some laws of physics, in the form of governing equations (see (Champion et al. 2019)). For that, the obtained model needs to be *interpretable* – that is, capable of explaining why data behave in a certain manner. An interpretable model has, in general, not only good generalization capabilities, but also good expressive ability – extrapolation (Kutz and Brunton 2022).

Think to the example in section 2.5 involving data generated from a combination of trigonometric functions with additive noise. The example highlighted that regression is actually more nuanced than simply choosing a regression model and performing a least-square fit. The selection of the model itself is crucial for achieving better predictions and interpretable descriptions of the data.

In the example, we have seen that if we pick an "optimal" polynomial degree, then we will obtain a model able to interpolate quite well the test data (see Figure 2.12, left). However, if the polynomial degree is too large (or too small), we will end up having a model that massively overfits (or underfits) the training data – that is, unable to generalize for new data points (see Figure 2.12, right). The ultimate solution was to resort to regularized regression models (e.g. LASSO). In so doing, we succeeded in identifying a good approximation of the true (trigonometric) model hidden under the pile of noisy data (see Figure 2.17).

However, why does a regularized regression model seem to have much more predictive power than a straightforward nonlinear regression one, with as many polynomial terms as we are able to count? The answer lies in Occam's razor, a principle of parsimony ("*lex parsimoniae*") attributed to William of Occam:

"*Entia non sunt multiplicanda praeter necessitatem*".

That is, "entities must not be multiplied beyond necessity". In other words, among different competing models that make the same predictions, the simpler one is the more likely. This principle is extensively used in science, where simpler explanations are preferred. An example is the method of dominant balance, which allows to determine the asymptotic behavior of solutions to an ordinary differential equation by identifying those terms that may be neglected in the limit. Dominant balance is, for instance, when we consider the evolution of shear stress in a sand sample, and we do not model either its quantum state or the effects of the relativistic warping of space-time caused by the grains.

Regularized regression adheres to this principle by discouraging the inclusion of numerous polynomial terms – or, in general, of features – thus promoting model simplicity. In addition, regularization helps in avoiding overfitting and results in a high degree of stability. Indeed, high-dimensional polynomial models tend to suffer from instability due to multicollinearity that occurs when two or more independent variables have a high correlation with one another (see Figure 2.12, right).

While some of these benefits could also be obtained with low-degree polynomial models without the need of resorting to regularization, these models tend to underfit the training data and exhibit insufficient expressive power. It is also worth mentioning that successful ML models often exhibit a certain degree of *benign overfitting* (Bartlett et al. 2020, also called *overparametrization*). Therefore, regularized regression models, with their ability to balance complexity and simplicity, offer significant predictive power by selecting relevant features.

In the example involving LASSO, Ridge and Elastic Net regression (section 2.6.2), we demonstrated the interpolating power of regularized regression models. However, it is important to further assess their interpretability and potential to provide insights for the identification and discovery of patterns and equations hidden within data sets.

2.7.1. *Interpretable models and where to find them*

This application deals with the investigation of whether it is possible to discover interpretable models (governing equations) with extrapolative power using (regularized) regression methods. In particular, we mimic virtual experiments governed by a very simple equation and consider additive noise simulating virtual statistical errors in a hypothetical data acquisition system. The underlying governing equation is the (one-dimensional) projectile motion – that is, the solution of the second-order ordinary differential equation $\ddot{x}(t) + g = 0$:

$$x(t) = x(0) + \dot{x}(0)t - \frac{1}{2}gt^2,$$

where x, $\dot{x}$ and $\ddot{x}$ are the trajectory, velocity and acceleration of the projectile, respectively; t is the time and g is the gravitational acceleration. The code hereinafter integrates the differential equation with initial conditions: $x(0) = 0$ m and $\dot{x} = 5$ m/s. As we are interested in making predictions within the range of values of some training data and outside of it, we generate the data and split them in an interpolation (`xi` and `yi`) and extrapolation (`xe` and `ye`) sets, as depicted in Figure 2.18(a):

```
import numpy as np
from scipy.integrate import solve_ivp
np.random.seed(42)

def projectile_motion(t, y):
    # equations of motion
    dydt = y[1]  # velocity
    dvdt = -g    # acceleration
    return [dydt, dvdt]
# Initial conditions
n_snapshots = 200
x0 = 0.0  # meters
v0 = 5.0  # meters per second
g = 9.81  # acceleration due to gravity
t0 = 0.0 # initial t
tf = 1.0 # final t
# time points solution
t = np.linspace(t0, tf, n_snapshots)
# solve ODE
sol = solve_ivp(projectile_motion, [t0, tf], [x0, v0], t_eval=t)
# define independent and dependent variables
X = np.float32(t)
y = np.float32(sol.y[0])
```

```
n = n_snapshots//2
xi = X[:n] # train time
xe = X[n:2*n] # test time
yi = y[:n] # train x = [0,0.5]
ye = y[n:2*n]  # test x = [0.5,1]
```

We continue by training a set of n polynomial regression models of degree 20. Every model is trained against dependent variables lying between 0 and 0.5 (xi) and dependent variables (yi) with a normally distributed measurement noise, randomly selected at each loop (fni). In a nested for loop, we store the MSE values between the models predictions and yi for values of the independent variables within the interpolation (mse_int) and extrapolation (mse_ext) range.

```
from sklearn.linear_model import LinearRegression
from sklearn.preprocessing import PolynomialFeatures
from sklearn.metrics import mean_squared_error
degree = 20 # polynomial
mse_int = np.zeros((n,degree))
mse_ext = np.zeros((n,degree))
poly_features = PolynomialFeatures(degree=degree,
                                   include_bias=False)
xi_poly = poly_features.fit_transform(np.expand_dims(xi,1))
xe_poly = poly_features.fit_transform(np.expand_dims(xe,1))
for deg in range(degree):
    xi_poly_ = xi_poly[:,:deg+1]
    xe_poly_ = xe_poly[:,:deg+1]
    for j in range(n):
        fni = yi + 0.3*np.random.normal(0,1,n)
                                   # y with normal noise
        lin_reg = LinearRegression() # model
        lin_reg.fit(xi_poly_,fni) # fit for x in [-0.5,1]
        ynai = lin_reg.predict(xi_poly_)
        ynae = lin_reg.predict(xe_poly_) # fit for t in [0.5,1]
        mse_int[j,deg] = mean_squared_error(yi,ynai) # MSE train
        mse_ext[j,deg] = mean_squared_error(ye,ynae) # MSE val
```

Figure 2.18 shows the MSE of the set of models in the interpolation and extrapolation ranges. The results are clear: all the trained models massively overfit the data with measurement noise and cannot generalize. When attempting to extrapolate beyond the range observed in the training data, we are confronted with substantial errors, as shown in Figures 2.18(c) and (d). The logarithmic plot in Figure 2.18(d) illustrates the exponential growth of the error as the polynomial degree increases, reaching magnitudes as high as 10^{10}. This demonstrates the inability of overfitted models to make predictions in the extrapolation range. Thus, we can say that (non-regularized) polynomial regression does not offer an interpretable model for this simple case (except an obvious second-order polynomial).

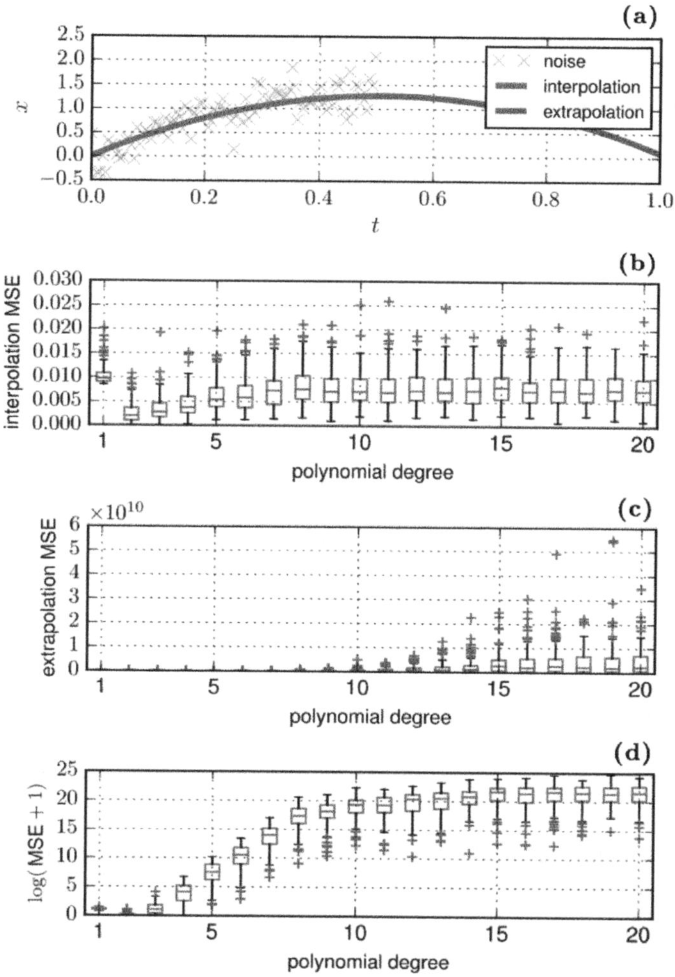

Figure 2.18. *Regression of the motion of a projectile with measurement noise (a) using 100 different nonlinear regression models with polynomial basis (degree 20). Errors for values within the training, interpolation range (b) and outside (extrapolation) (c and d)*

Let us continue by employing cross-validation to mitigate the effects of overfitting. We also extend the set of polynomial regression models – denoted with pinv, diminutive of pseudoinverse – with regularized ones. The code below performs 100-fold cross-validation for the same noisy data sets for the three different regression models (pinv, LASSO and Ridge) and stores the obtained parameters. Note that, as the measurement noise is randomly select for each k-fold, the training data points may be considered as virtual experimental measurements of the simplified system.

```
from sklearn.linear_model import Lasso,Ridge
from sklearn.model_selection import KFold
nfolds = 100
lin_reg = LinearRegression() # linear model
lasso = Lasso(alpha=3e-4) # LASSO model
ridge = Ridge(alpha=3e-4) # Ridge model
k_fold = KFold(nfolds)
# initialize for storing parameters
lasso_params = np.zeros((nfolds,degree+1))
ridge_params = np.zeros((nfolds,degree+1))
lin_params = np.zeros((nfolds,degree+1))
for k, (train, test) in enumerate(k_fold.split(xi_poly, yi)):
    fni = yi + 0.3*np.random.normal(0,1,n) # y with normal noise
    lin_reg.fit(xi_poly[train], fni[train]) # fit lin
    ridge.fit(xi_poly[train], fni[train]) # fit RIdge
    lasso.fit(xi_poly[train], fni[train]) # fit LASSO
    # store models parameters
    lin_params[k,0] = lin_reg.intercept_
    lin_params[k,1:] = lin_reg.coef_
    ridge_params[k,0] = ridge.intercept_
    ridge_params[k,1:] = ridge.coef_
    lasso_params[k,0] = lasso.intercept_
    lasso_params[k,1:] = lasso.coef_
```

Figure 2.19 plots the values of the parameters $\boldsymbol{\theta}$ obtained from the above code for kth fold ($k = 2$, 10 and 100). Note how the parameters of the standard regression model explode, with values of the order of 10^5, independently of the number of folds. The phenomenon is intrinsic to polynomial regression that blows up as the polynomial degree is increased and further enhanced by the intrinsic multicollinearity of the (polynomial) features. In contrast, Ridge and LASSO identify parameters with ranges of values comparable with underlying equation of motion. In addition, LASSO also successfully identifies a parsimonious model, with a (more or less) predominant quadratic term (depending on the number of folds).

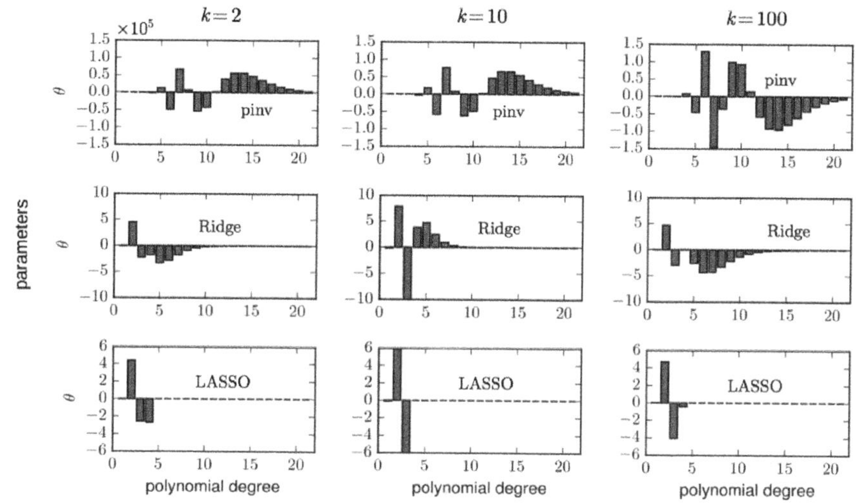

Figure 2.19. *Values of the parameters θ obtained using standard (pinv), LASSO and Ridge polynomial regression for different cross-validation folds:* $k = 2$ *(left),* $k = 10$ *(center) and* $k = 100$ *(right)*

In order to proceed to the evaluation of the three cross-validated models, we compute the average of the parameters obtained from the optimization performed at each fold (see section 2.5.1.2):

```
# Compute models based on the mean over the
                              100-fold cross-validation
lin_theta = np.mean(lin_params[:,:],0)
lasso_theta = np.mean(lasso_params[:,:],0)
ridge_theta = np.mean(ridge_params[:,:],0)
# Set mean params for each model
lin_reg.coef_ = lin_theta[1:]
lin_reg.intercept_ = lin_theta[0]
ridge.coef_ = ridge_theta[1:]
ridge.intercept_ = ridge_theta[0]
lasso.coef_ = lasso_theta[1:]
lasso.intercept_ = lasso_theta[0]
```

```
# Predictions within interpolation range
fy_lin_xi = lin_reg.predict(xi_poly)
fy_la_xi = lasso.predict(xi_poly)
fy_rid_xi = ridge.predict(xi_poly)
# Predictions within extrapolation range
fy_la_xe = lasso.predict(xe_poly)
fy_lin_xe = lin_reg.predict(xe_poly)
fy_rid_xe = ridge.predict(xe_poly)
```

Figure 2.20 compares the performances of the three models (pinv, LASSO and Ridge) in terms of the mean squared error, while Figure 2.21 shows the predictions in the interpolation and extrapolation range. As we might have expected, LASSO excels over all other strategies by providing a model that can interpolate (without overfitting) and, even, extrapolate with good accuracy. We can also observe that Ridge performs better than the standard regression approach, although it rapidly fails in the extrapolation range.

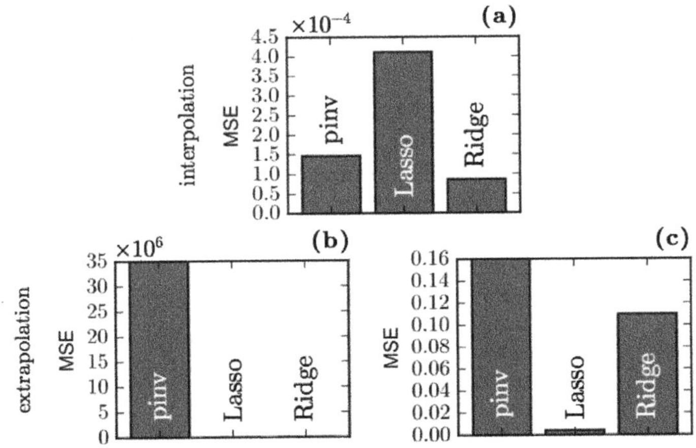

Figure 2.20. *Errors of the cross-validated regression models: pinv, LASSO and Ridge within the (a) interpolation range and (b and c) extrapolation range, where (c) presents a detailed view*

With good generalization and extrapolation capabilities, LASSO has not identified the exact underlying governing equation but just a very good approximation of it:

$$\hat{x}(t) \approx \theta_1 t + \theta_2 t^2 + \theta_3 t^3, \quad \theta_1 = 4.33, \quad \theta_2 = -2.93, \quad \theta_3 = 1.32.$$

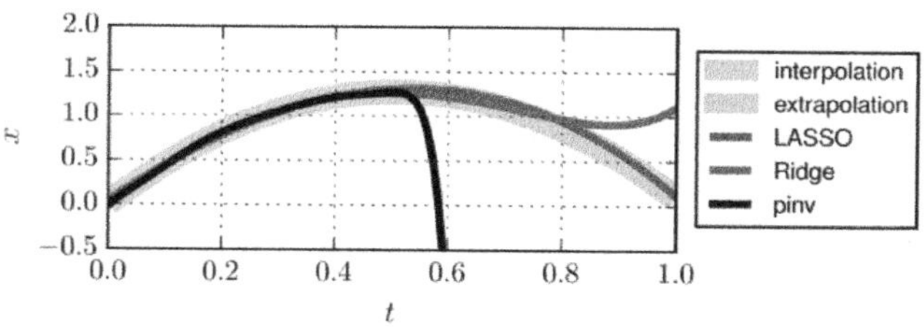

Figure 2.21. *Comparison between the projectile motion equation and the predictions of the cross-validated models: pinv, LASSO and Ridge*

Note that the true equation has $\theta_1 = 5$, $\theta_2 = -4.9$ and $\theta_3 = 0$. This implies that if we ask the model to make predictions for a significantly distant time point, that is, $t \gg 1$, the accuracy of the extrapolation will not be as good as in the range that we considered here.

2.8. Bayesian regression

All regression methods discussed up to this point fall under the *frequentist approaches* (Wakefield 2013) as they assume that there are enough measurements to say something meaningful about $\boldsymbol{\theta}$ and provide a single best estimate for a given training set. Such methods provide good models whenever the relationship between the independent and dependent variables is well understood and relatively stable. However, in real-case scenarios, we are often confronted with uncertainty or variability in the data and deterministic methods are not appropriate. To this end, a powerful, alternative tool is provided by Bayesian regression methods.

Bayesian regression is able to provide predictions in the form of probability distributions incorporating uncertainty estimates. This section introduces the main ideas and novelties characterizing such methodologies as compared to frequentist approaches by focusing on linear Bayesian regression and GP regression. For more details, the reader can refer to (Bishop and Nasrabadi 2006; Murphy 2012).

2.8.1. *Linear Bayesian regression*

In simple linear regression, we aim to find the best-fit line that explains the relationship between two variables. Linear Bayesian regression extends this same concept by considering uncertainty in both the model parameters and the predictions. Instead of relying on a single line (see section 2.2), it thus provides a distribution of possible lines, accounting for the uncertainty in the data.

Linear Bayesian regression starts by assuming a linear relationship between the independent variables $\mathbf{X}$ and the dependent variable $\mathbf{y}$ of the form:

$$\mathbf{y} = \mathbf{X}\boldsymbol{\theta} + \epsilon, \qquad [2.27]$$

where $\boldsymbol{\theta}$ are the model parameters and ϵ represents an error term assumed to follow a zero-mean Gaussian distribution, $\epsilon \sim \mathcal{N}(\mathbf{0}, \sigma^2\mathbf{I})$, where $\mathbf{I}$ is the identity matrix and σ^2 is the variance.

2.8.1.1. *Bayesian inference*

Bayesian inference allows to infer a posterior distribution $p(\boldsymbol{\theta}|\mathbf{X}, \mathbf{y})$ – that of the parameters – representing updated beliefs given the observed independent variables $\mathbf{X}$ and dependent variable $\mathbf{y}$. To this end, we use Bayes' formula, that is:

$$p(\boldsymbol{\theta}|\mathbf{X}, \mathbf{y}) \propto p(\mathbf{X}, \mathbf{y}|\boldsymbol{\theta})\, p(\boldsymbol{\theta}), \qquad [2.28]$$

where $p(\boldsymbol{\theta})$ is the prior distribution over $\boldsymbol{\theta}$, incorporating prior beliefs about the parameters values before observing any data, while $p(\mathbf{y}|\mathbf{X}, \boldsymbol{\theta})$ is the likelihood function, quantifying instead the probability of observing the dependent variable $\mathbf{y}$ given the independent variables $\mathbf{X}$ and the parameters $\boldsymbol{\theta}$.

2.8.1.2. *Prior and posterior*

In principle, any prior $p(\boldsymbol{\theta})$ could be used, however the functional form of most priors, when multiplied by the functional form of the likelihood in [2.28], results in an posterior $p(\boldsymbol{\theta}|\mathbf{X}, \mathbf{y})$ with no closed-form solution.

In such scenarios, the solution is to resort to approximate Bayesian inference techniques as Monte Carlo methods. But, certain priors are mathematically convenient because they result in posteriors with tractable, well-known densities.

The simplest and most widely used version of linear Bayesian regression is the *normal linear model*, in which $\mathbf{y}$ given $\mathbf{X}$ – $p(\mathbf{X}, \mathbf{y}|\boldsymbol{\theta})$ – is distributed Gaussian. In this model, and under a particular choice of the prior for the parameters – namely, when the prior is conjugate, meaning that the prior and posterior share the same functional form – the posterior can be found analytically. With more arbitrarily chosen priors, the posteriors generally have to be approximated.

2.8.1.3. *Posterior predictive*

Once we have computed the posterior distribution $p(\boldsymbol{\theta}|\mathbf{X}, \mathbf{y})$, we can make predictions for new, unseen data points. One common approach is to generate posterior predictive samples by drawing parameter values from the posterior distribution and using them to make predictions (Bishop and Nasrabadi 2006).

2.8.1.4. *Example*

Let us consider an intuitive example drawn from (Bishop and Nasrabadi 2006) and implemented as in https://github.com/gwgundersen/bayesian-linear-regression. We start by generating 50 snapshots of the independent and dependent variables, according to:

$$y = \theta_1 \mathbf{x} + \theta_0,$$

where $\theta_1 = 0.5$ and the bias is $\theta_0 = -0.7$. Then, we perform Bayesian linear regression using as prior a normal-inverse-gamma distribution and consider an increasing number of observations in the training data set.

Figure 2.22 depicts the evolution of the prior and posterior distributions of the parameters $\boldsymbol{\theta}$ (left column) and six random posterior samples drawn from the vector $\boldsymbol{\theta} = [\theta_0, \theta_1]^{\mathrm{T}}$ (right column). In the top row, the model has seen no data. The prior places high and equal probability on both θ_0 and θ_1 being zero. In the subsequent rows, the model is fit to more data. With more observations, the model inferred posterior variance decreases, and the realizations of $\boldsymbol{\theta}$ from the posterior become more constrained and in agreement with the observed data.

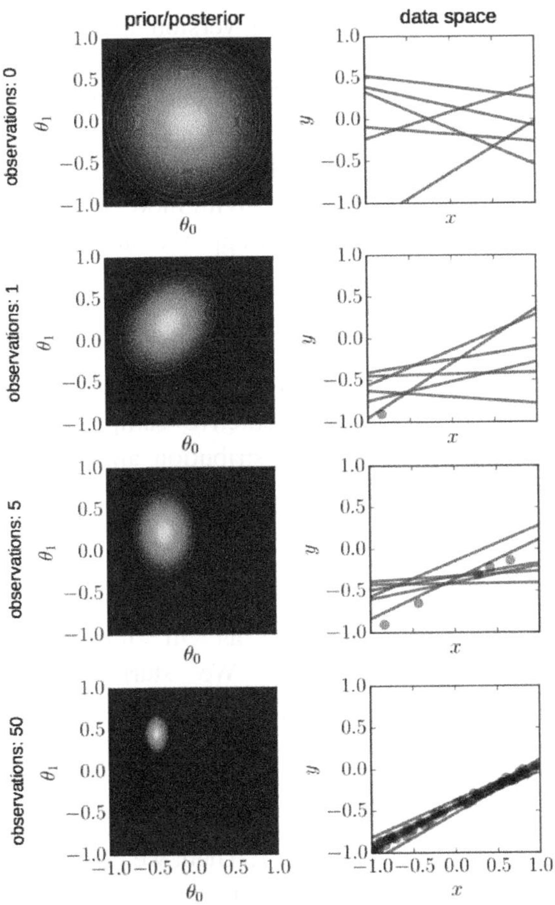

Figure 2.22. *Evolution of the prior and posterior distributions of the parameters $\boldsymbol{\theta}$ (left column) and random samples of the vector $\boldsymbol{\theta} = [\theta_0, \theta_1]^T$ (right column) at varying of the number of observation (rows)*

2.8.2. *GP regression*

Before discussing GP regression, let's distinguish *parametric* from *non-parametric* models and recall the definition of a GP.

2.8.2.1. *Parametric versus non-parametric model*

Parametric models are all those regression models we have seen so far. More formally, parametric models assume a well-defined functional relationship between dependent and independent variables and that the distribution of the data can be entirely defined in terms of a finite set of parameters $\boldsymbol{\theta}$. Thus, once we have trained a parametric model, every future prediction is independent of the particular set of new data for which we are making predictions.

Non-parametric models do not prescribe any predetermined functional relationship (or hypothesis $f_{\boldsymbol{\theta}}$) between the dependent and independent variables (Hardle and Mammen 1993). Non-parametric models thus have the freedom to calculate the probability distribution over all admissible functions that fit the data, rather than, for instance, calculating the probability distribution of parameters of a specific function (see section 2.8.1). However, even in non-parametric models, we must specify a prior (on the function space), calculate the posterior using the training data and compute the predictive posterior distribution on the points of interest.

2.8.2.2. *Gaussian process*

A GP is a collection of random variables, any finite number of which has a joint Gaussian distribution. In the context of regression, we can think of a GP as an infinite-dimensional generalization of a multivariate Gaussian distribution. Instead of representing a mean vector and a covariance matrix, a GP is fully characterized by a mean function, denoted as $\mu(\mathbf{x})$, and a covariance function, also known as the kernel function, denoted as $k(\mathbf{x}, \mathbf{x})$, describing the relationship between any given two data points $\mathbf{x}_i$ and $\mathbf{x}_j$, for $i, j = 1, 2, \cdots, m$.

2.8.2.3. *Inference*

Gaussian process regression (GPR) is a non-parametric, Bayesian regression method. In particular, GPR uses a GP prior to infer the distribution of possible functions that could generate the observed data.

To fix the ideas, consider a set of training data $\{\mathbf{X}, \mathbf{y}\} \equiv \{(\mathbf{x}^{(i)}, y^{(i)})\}_{i=1}^{n}$; the objective is to estimate the function $f(\mathbf{X})$ that minimizes $\mathbf{y} - f(\mathbf{X}) = 0$ by placing a GP over it. Typically, a mean of zero is assumed, leading to the following distribution:

$$\mathbf{y} = f(\mathbf{X}) \sim \mathcal{N}(\mathbf{0}, \mathbf{K}), \qquad [2.29]$$

where $\mathbf{K} \equiv k(\mathbf{X}, \mathbf{X})$ denotes the covariance matrix generated by the kernel function k chosen (e.g. periodic, linear and radial basis function) and describes the general shape of $f(\cdot)$. Note that the choice of the kernel is crucial in identifying a good (or a bad) fit of the data.

To predict the values of a new set of dependent variables $\mathbf{y}_*$, given a new set of independent variables $\mathbf{X}_*$, we need to estimate the conditional distribution $p(\mathbf{y}_*|\mathbf{X}_*, \mathbf{X})$. According to the properties of GP (Murphy 2012), the joint distribution of the training and test outputs follows a joint Gaussian distribution:

$$\begin{bmatrix} \mathbf{y} \\ \mathbf{y}_* \end{bmatrix} = f\left(\begin{bmatrix} \mathbf{x} \\ \mathbf{x}_* \end{bmatrix}\right) \sim \mathcal{N}\left(\mathbf{0}, \begin{bmatrix} \mathbf{K} & \mathbf{K}_* \\ \mathbf{K}_*^{\mathrm{T}} & \mathbf{K}_{**} \end{bmatrix}\right), \qquad [2.30]$$

where $\mathbf{K}_* \equiv k(\mathbf{X}, \mathbf{x}_*)$, $\mathbf{K}_{**} \equiv k(\mathbf{x}_*, \mathbf{x}_*)$. While the above joint distribution gives some insight as to how $f(\mathbf{x}_*)$ relates to $f(\mathbf{x})$, at this point no prediction for the new datum $\mathbf{x}_*$ is made.

To obtain the posterior distribution of the predicted GP realizations $f(\mathbf{x}_*)$, we condition the prior distribution on the training data (Murphy 2012). This leads to the following mean and covariance predictions for the test point $\mathbf{x}_*$ (see Murphy (2012), for the detailed derivation):

$$\mu\big(f(\mathbf{x}_*)\big) = \mathbf{K}_*^{\mathrm{T}}\mathbf{K}^{-1}\mathbf{y}, \qquad [2.30a]$$

$$\mathrm{Cov}\big(f(\mathbf{x}_*)\big) = \mathbf{K}_{**} - \mathbf{K}_*\mathbf{K}^{-1}\mathbf{K}_*^{\mathrm{T}}. \qquad [2.30b]$$

These equations provide the mean and covariance predictions for the output $f(\mathbf{x}_*)$ given the training data set and a new input $\mathbf{x}_*$.

2.8.2.4. *Inference in the presence of noise*

In real-world scenarios, we often encounter observations with measurement noise. In such scenarios, it is more appropriate to model the training targets $\mathbf{y}$ to be noisy realizations of a Gaussian process $f(\mathbf{X})$:

$$\mathbf{y} = f(\mathbf{X}) \sim \mathcal{N}(\mathbf{0}, \mathbf{K}) + \epsilon, \quad \epsilon \sim \mathcal{N}(0, \sigma^2), \qquad [2.31]$$

where the noise ϵ is parameterized by a zero-mean Gaussian with positive noise covariance values given by σ^2, which is a hyperparameter.

By repeating the steps above, we obtain the following joint distribution of the training and new outputs:

$$f\left(\begin{bmatrix} \mathbf{x} \\ \mathbf{x}_* \end{bmatrix}\right) \sim \mathcal{N}\left(\mathbf{0}, \begin{bmatrix} \mathbf{K} + \sigma^2\mathbf{I} & \mathbf{K}_* \\ \mathbf{K}_*^{\mathrm{T}} & \mathbf{K}_{**} \end{bmatrix}\right). \qquad [2.32]$$

As before, but now with the noise term, we can obtain the predictive mean and variance, given by:

$$\mu(f(\mathbf{x}_*)) = \mathbf{K}_*^{\mathrm{T}} \left(\mathbf{K} + \sigma^2\mathbf{I}\right)^{-1} \mathbf{y}, \qquad [2.33a]$$

$$\mathrm{Cov}(f(\mathbf{x}_*)) = \mathbf{K}_{**} - \mathbf{K}_* \left(\mathbf{K} + \sigma^2\mathbf{I}\right)^{-1} \mathbf{K}_*^{\mathrm{T}} + \sigma^2. \qquad [2.33b]$$

2.8.2.5. *Example*

Let us see how we can perform GRP using Scikit-Learn. We consider the same polynomial regression example in section 2.5, with data generated from a combination of trigonometric functions with random additive noise. In the code hereinafter, we first fit a GP on the training data using a radial basis function (`RBF`) kernel and, additionally, a noise parameter (`alpha`). Then, we use the kernel to compute the mean predictions (`norm_y_mean`) for the train and test set and plot the 95% confidence interval (defined as `norm_y_mean` ± 1.96 `norm_y_std`).

```
from sklearn.gaussian_process import GaussianProcessRegressor
from sklearn.gaussian_process.kernels import RBF
# define a radial basis function kernel
kernel = 1 * RBF(length_scale=1.0,
                 length_scale_bounds=(0.001, 10)) #
GPR = GaussianProcessRegressor(kernel=kernel,
                               alpha=0.5**2) # noise term
GPR.fit(norm_X_train, norm_y_train)
                               # find mean and covariance matrix
# posterior distribution
norm_y_mean, norm_y_std = GPR.predict(norm_x, return_std=True)
                               # see par. 5.1
```

The predictions are depicted in Figure 2.23 and compared with the training and test data. Not only the fit in terms of the expected (mean) value is extremely good, but in addition, due to the intrinsic probabilistic nature of GPR, we are also able to compute confidence intervals and thus quantify uncertainties.

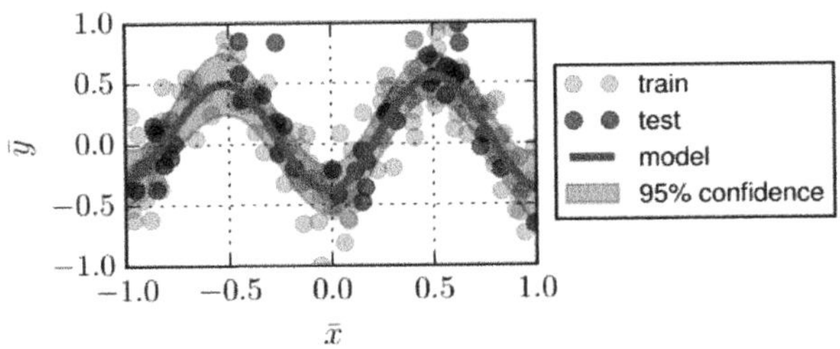

Figure 2.23. *Gaussian process regression (with noise) for data generated from a combination of trigonometric functions with random additive noise (see section 2.5.1): predictions and 95% confidence interval*

In summary, we have seen that Bayesian regression methods offer unique advantages as compared to classical, frequentist ones as far as it concerns uncertainty estimation. However, there exist some drawbacks associated with the former. First, both linear Bayesian and GP regression have high computational complexity – higher than that of deterministic methods – and may be computationally expensive, especially when dealing with large data sets and not only at training but also at inference. Second, we have seen that deterministic regression methods allow to identify interpretable models that directly relate to the impact of each parameter on the dependent variable predictions – especially in the case of regularized strategies (see section 2.7). However, this is not usually true for linear Bayesian and GP regression: indeed, the parameters are expressed in the form of posterior distributions or covariance matrices and their interpretation can be quite challenging.

2.9. Conclusions

This chapter provided a comprehensive introduction to regression methods, which serve as one of the foundational pillars of ML, together with classification methods (see Chapters 1 and 4).

We covered linear and nonlinear regression models and their Bayesian counterparts in sections 2.2, 2.5 and 2.8, respectively. Together, such statistical tools allow to identify relationships between inputs and outputs and deliver predictive models. To do so, the starting point consists of defining a hypothesis – in linear and nonlinear regression – or a kernel – in GP regression. Then, the process continues with the identification of the hypothesis parameters – in the form of fixed values or posterior distributions – or functionals, obtained through optimization strategies (GD) or as closed-form solutions, to finally obtain a predictive model through learning from a training data set.

We also introduced concepts and strategies related to model validation and generalization, as well as features scaling (see section 2.4). These are extremely important and useful not only for regression methods but, in general, for any ML approach – for example, classification methods, artificial neural networks and dimensionality reduction techniques.

Then, we emphasized the significance of regularization, particularly in the context of nonlinear regression (see section 2.6). Regularization techniques play a crucial role in preventing overfitting, where the model becomes too complex and fits noise or irrelevant patterns, leading to poor generalization. Note that regularization strategies go well beyond regression models and are a good ally for enhancing the performance and generalization abilities of

ML models. Additionally, regularization facilitates the identification of interpretable models by promoting simplicity and a small number of dominant parameters. Interpretable models have the potential to reveal hidden physical relationships or equations that may be obscured within vast amounts of data (see section 2.7).

However, we should pay attention in understanding the limitations of (non)-regularized regression models when addressing complex phenomena. For instance, if we were asked to predict occurrences and magnitudes of earthquakes and landslides or characterize the constitutive behavior of intricate materials like sand or clay, the interpretable, but simplistic, approach developed in the case of the projectile motion (see section 2.7.1) will not be quite effective. The reason lies in challenges proper to the above scenarios and related to an intrinsic chaotic nature and a high-dimensional state space.

Every time we are confronted with the forecasting of a complex phenomenon or the description of an intricate system achieving reliable generalization and extrapolation, performance becomes increasingly arduous when relying solely on ML approaches that lack any knowledge bias. And this even in the case of sophisticated neural networks that use regularization strategies (see Chapter 5). To address such challenges, it becomes imperative to resort to physics-informed and thermodynamics-based ML models (as explored in Chapters 2 and 3 of volume 2) or leverage data-driven computing approaches (presented in Chapter 1 of volume 2). These alternative methodologies incorporate domain knowledge and physical principles into the learning process, resulting in more robust and accurate predictions.

2.10. References

Bartlett, P.L., Long, P.M., Lugosi, G., Tsigler, A. (2020). Benign overfitting in linear regression. *Proceedings of the National Academy of Sciences*, 117(48), 30063–30070.

Bishop, C.M. and Nasrabadi, N.M. (2006). *Pattern Recognition and Machine Learning*, vol. 4, Springer, Singapore.

Boyd, S.P. and Vandenberghe, L. (2004). *Convex Optimization*. Cambridge University Press, Cambridge.

Buitinck, L., Louppe, G., Blondel, M., Pedregosa, F., Mueller, A., Grisel, O., Niculae, V., Prettenhofer, P., Gramfort, A., Grobler, J. et al. (2013). API design for machine learning software: Experiences from the scikit-learn project. *arXiv*. doi: 10.48550/arXiv.1309.0238.

Champion, K., Lusch, B., Kutz, J.N., Brunton, S.L. (2019). Data-driven discovery of coordinates and governing equations. *Proceedings of the National Academy of Sciences*, 116(45), 22445–22451.

Gentle, J.E. (2012). *Numerical Linear Algebra for Applications in Statistics*. Springer Science & Business Media, New York.

Géron, A. (2022). *Hands-On Machine Learning with Scikit-Learn, Keras, and TensorFlow*. O'Reilly Media, Inc, Sebastopol, CA.

Hardle, W. and Mammen, E. (1993). Comparing nonparametric versus parametric regression fits. *The Annals of Statistics*, 21(4), 1926–1947.

Hoerl, A.E. and Kennard, R.W. (1970). Ridge regression: Biased estimation for nonorthogonal problems. *Technometrics*, 12(1), 55–67.

Jiang, Y., Neyshabur, B., Mobahi, H., Krishnan, D., Bengio, S. (2019). Fantastic generalization measures and where to find them. *arXiv*. doi: 10.48550/arXiv.1912.02178.

Kutz, J.N. and Brunton, S.L. (2022). Parsimony as the ultimate regularizer for physics-informed machine learning. *Nonlinear Dynamics*, 107(3), 1801–1817.

Mandel, J. (1982). Use of the singular value decomposition in regression analysis. *The American Statistician*, 36(1), 15–24.

Murphy, K.P. (2012). *Machine Learning: A Probabilistic Perspective*. MIT Press, Cambridge, MA.

Tibshirani, R. (1996). Regression shrinkage and selection via the lasso. *Journal of the Royal Statistical Society: Series B (Methodological)*, 58(1), 267–288.

Wakefield, J. (2013). *Bayesian and Frequentist Regression Methods*, vol. 23. Springer, New York.

Wichtmann, T. and Triantafyllidis, T. (2016). An experimental database for the development, calibration and verification of constitutive models for sand with focus to cyclic loading: Part I – Tests with monotonic loading and stress cycles. *Acta Geotechnica*, 11, 739–761.

3

Unsupervised Learning: Basic Concepts and Application to Particle Dynamics

Noel JAKSE
SIMaP, CNRS, Grenoble INP, Université Grenoble Alpes, France

3.1. Introduction

Machine learning (ML) approaches attracted significant interest in many scientific fields due to their potential ability to uncover patterns, make predictions and extract valuable insights from large and complex body of data (Efron and Hastie 2021). Web-based data collection platforms, scientific instruments and computer simulations are creating exponentially increasing data stores. They triggered new scientific methods to analyze and organize huge amount of data, giving the

---

For a color version of all the figures in this chapter, see www.iste.co.uk/stefanou/machine1.zip.

Machine Learning in Geomechanics 1,
coordinated by Ioannis STEFANOU and Félix DARVE.

possibility to find subtle effects missed previously (Hey et al. 2009; Hey and Trefethen 2020).

ML tools can be broadly categorized into three types: supervised, unsupervised and semi-supervised learning. Each type offers distinct approaches to learning from data, with specific strengths and limitations (Efron and Hastie 2021; Hastie et al. 2009). Supervised learning is one of the most used approachs and involves training a model using labeled data, where both input features and corresponding output labels are provided. The model learns to map inputs to their respective outputs, generalizing to make predictions for new, unseen data. Applications of supervised learning include regression and classification tasks that will be the subject of Chapters 2 and 4.

However, obtaining labeled data can be time consuming and expensive, especially for large datasets. Unsupervised learning, on the other hand, works with unlabeled data, where the model discovers patterns and structures within the data without prior knowledge of the desired output. It is particularly useful in exploratory data analysis, clustering, dimensionality reduction and anomaly detection. Finally, semi-supervised learning combines aspects of both supervised and unsupervised learning, leveraging a small amount of labeled data with a large amount of unlabeled data. The method is based on unsupervised techniques to extract features from the unlabeled data, and supervised learning is then applied to improve the trained model using the labeled data. This approach often results in improved model performance compared to purely supervised or unsupervised learning, especially when labeled data are limited.

Unsupervised learning, at the heart of the present chapter, is useful in situations where labeled data are scarce, enabling insights from larger datasets. It can be useful to deal with large datasets of unlabeled instance, and is particularly relevant when hidden patterns and structures in data have to be extracted without prior knowledge on any label or output. It plays a vital role in scientific fields such as materials science (Ceriotti 2019; Schmidt et al. 2019) and geophysics (Walker and Tordesillas 2010; Karapiperis and Andrade 2021), where labeled data are often unavailable, scarce or expensive to calculate or measure.

It enables to analyze large datasets and identify essential relationships, leading to the discovery of new materials or the understanding of complex geological processes.

The chapter is mainly divided into two parts. The first part is devoted to the comprehensive description of the basic concepts and most popular techniques of unsupervised learning. For the latter, the choice was made to describe two main branches, which are the clustering and dimensionality reduction techniques (Hastie et al. 2009). Unsupervised learning has numerous real-world applications in many domains. As an example of applications, the second part of this chapter illustrates an application of unsupervised learning to the discovery of patterns in particles dynamics. A particular focus is made on large-scale molecular dynamics (MD) simulations performed with up to 10 million atoms for the purpose of describing the early stages of solidification of a material, called homogeneous nucleation (Sosso et al. 2016; Becker et al. 2022a). While specific to material science, this example can surely be transposed at higher scales with discrete particle dynamics (DPD) modeling in geophysics, which is at the heart of this book.

3.2. Basic concepts

3.2.1. *Representation of the data: Feature extraction and selection*

Feature extraction and representation are fundamental aspects of unsupervised learning as in most of the ML techniques. The basic idea sketched in Figure 3.1 is to transform raw data into a format, often a vector, that can be easily processed and analyzed by ML algorithms; such a representation is so-called hereafter a descriptor. Having said that, the major objective of this transformation is then to identify and extract the most relevant and informative features from the data, intending to capture the underlying patterns and structures while reducing noise and redundancy.

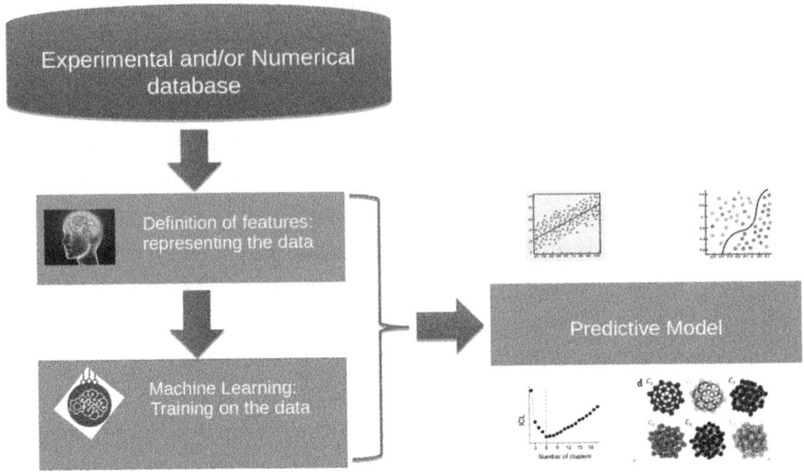

Figure 3.1. *Typical machine learning flowchart that learns a predictive model from the data*

Feature extraction techniques can be divided into two categories: feature construction and feature selection (Khalid et al. 2014). As mentioned above, feature construction involves creating new features from the original data, often through mathematical transformations or domain-specific knowledge. For example, in image processing, features can be extracted using edge detection or texture analysis. In text analysis, natural language processing techniques such as tokenization, stemming and term frequency-inverse document frequency (TF-IDF) can be employed to extract meaningful features (Nothman et al. 2018) in the so-called Bag of Words representation. In the application treated in the present chapter, where atomic scale simulations of homogeneous crystal nucleation are considered, a topological descriptor is built (Becker et al. 2022b) based on persistence homology (PH) in the framework of the topological data analysis (TDA) (Pirashvili et al. 2018; Carlsson 2020).

This poses naturally the question of the dimensionality of the descriptor after extraction. High-dimensional feature spaces can pose challenges for unsupervised learning algorithms, as the increase in dimensions can lead to increased computational complexity, reduced model interpretability and over-fitting. This phenomenon is known as the "curse of dimensionality" (Hastie et al. 2009; Efron and Hastie 2021). Feature selection addresses this problem by selecting a subset of the original features that are most relevant to the task at hand in order to reduce the dimensionality of the data. There are three primary methods for feature selection among others. Filter methods evaluate the relevance of individual features based on their statistical properties, independent of any learning algorithm. Examples of filter methods include correlation coefficients, mutual information and chi-squared tests. Wrapper methods use a specific learning algorithm's performance to guide the feature selection process by iteratively adding or removing features based on their impact on the model's performance (i.e. forward selection, backward elimination and recursive feature elimination).

Finally, embedded methods integrate feature selection into the model training process, combining the advantages of both filter and wrapper methods. One advantage is the assessment of the importance of features during the training stage, automatically selecting the most relevant ones. Emblematic examples of embedded methods include Least Absolute Shrinkage and Selection Operator (LASSO), ridge regression and decision trees.

3.2.2. *Distance and similarity metrics*

In order to quantify the relationship between data points in the feature space of dimension D, defining metrics is at the heart of unsupervised learning. These metrics provide a basis for comparing and grouping similar instances in clustering.

The most widely used distance metric is measuring the straight-line distance between two points in Euclidean D-dimensional feature space. Considering two points p and q, it is given by:

$$d(p, q) = \sqrt{\sum_{i=1}^{D} (p_i - q_i)^2}. \qquad [3.1]$$

This applies for continuous features and is sensitive to the scale of the data. A simpler version is the Manhattan distance, also known as the L_1-norm, which measures the sum of the absolute differences, namely:

$$d_{L_1}(p, q) = \sum_{i=1}^{N} |p_i - q_i|, \qquad [3.2]$$

and is less sensitive to outliers compared to Euclidean distance. When the dimensionality of the feature space is high or when the magnitude of the vectors is less relevant, such as in text analysis, the cosine similarity metrics is relevant and measures the angle between two feature vectors $\mathbf{p}$ and $\mathbf{q}$ as:

$$\text{cosine similarity}(\mathbf{p}, \mathbf{q}) = \frac{\mathbf{p} \cdot \mathbf{q}}{\|\mathbf{p}\| \|\mathbf{q}\|} = \frac{\sum_{i=1}^{n} p_i q_i}{\sqrt{\sum_{i=1}^{n} p_i^2} \sqrt{\sum_{i=1}^{n} q_i^2}}. \qquad [3.3]$$

If sets of data have to be compared, the Jaccard similarity metrics can be used for comparing binary or set-based data, measuring the ratio of the intersection to the union of two sets A and B is given by:

$$J(A, B) = \frac{|A \cap B|}{|A \cup B|}. \qquad [3.4]$$

Alternatively, Pearson correlation coefficient measures the linear relationship between two variables of size n, ranging from -1 (perfect negative correlation) to 1 (perfect positive correlation) with:

$$r = \frac{\sum_{i=1}^{n} (X_i - \bar{X})(Y_i - \bar{Y})}{\sqrt{\sum_{i=1}^{n} (X_i - \bar{X})^2} \sqrt{\sum_{i=1}^{n} (Y_i - \bar{Y})^2}}. \qquad [3.5]$$

These are the most common ones, however the metrics to be used should be decided given the specificity of the data at hand. The choice can affect the results of the learning.

3.3. Unsupervised learning techniques

Unsupervised learning comprises a diverse range of methods and algorithms dealing with unlabeled data (Efron and Hastie 2021). This section provides an overview of some popular unsupervised learning techniques and their applications, focusing only on *clustering* and *dimensionality reduction* methods. A corresponding, comprehensive tutorial through a JUPYTER notebook provided in this chapter will explore the most common ones.

What will not be treated in the present chapter is the combination of powerful deep learning models with unsupervised learning techniques (Huang et al. 2013). These so-called *deep unsupervised techniques* are autoencoders and variational autoencoders (VAEs), generative adversarial networks (GANs), self-supervised learning and deep clustering to name the most common ones.

3.3.1. *Clustering*

Clustering is the grouping process of data points based on their features. Many clustering methods were already proposed (Hastie et al. 2009). The key notion is the degree of similarity (or dissimilarity) between the individual objects being associated. Indeed, the method attempts to group observations with respect to some similarity criterion. Combinatorial methods were proposed to avoid a modeling through a probability distribution function. However, an exact solution is computable only when the dataset is very small. As a matter of fact, testing every combination of 20 observations into four clusters needs 10^{10} comparisons, which is basically unreachable.

Approximations of these methods were proposed through iterative process. The famous K-means method is a standard combinatorial algorithm based on the Euclidean distance given by equation [3.1]. A standard version of the algorithm is given in algorithm 3.1. Given a number K of points in the dataset of size n, $X = \{x_1, x_2, \ldots, x_n\}$ are randomly chosen and denoted as the initial K centroids, and each observation is assigned to the cluster with the closest centroid. Then, the centroids are updated, and the process is iterated. As an illustration, Figure 3.2 schematizes the application of the K-means to a dataset randomly created as a collection of three distinct blobs[1] as a guide for the eyes (left panel), and with the same color to indicate the fact that the data are unlabeled. The left panel shows how the feature space is divided in three domains based on the shortest distance to the centroids of the three identified clusters. These domains highlight how the model will assign any new point to a given cluster.

Algorithm 3.1. K-means

Require: Data points $X = \{x_1, x_2, \ldots, x_n\}$, number of clusters K
Ensure: Cluster assignments $C = \{c_1, c_2, \ldots, c_K\}$, centroids $\mu = \{\mu_1, \mu_2, \ldots, \mu_K\}$
  Initialize centroids $\mu_1, \mu_2, \ldots, \mu_K$ randomly from X
  repeat
    Assign each point x_i to the nearest centroid: $c_i = \arg\min_k \|x_i - \mu_k\|^2$
    Update centroids by computing the mean of assigned points: $\mu_k = \frac{\sum_{i=1}^{n} I(c_i=k)x_i}{\sum_{i=1}^{n} I(c_i=k)}$
  until Convergence (centroids do not change significantly or a maximum number of iterations is reached)

Adding a probabilistic view on this modeling with a soft assignment of each observation to each cluster leads to Gaussian mixture models, with spherical covariance matrices, proportional to the identity matrix. The idea behind it is that the data distribution can be well approximated by a mixture of K multivariate Gaussian distributions:

$$\sum_{k=1}^{K} \pi_k \Phi(\mu_k, \Sigma_k), \qquad [3.6]$$

with Φ the multivariate Gaussian distribution, where the cluster $k \in \{1, \ldots, K\}$ is described by its proportion π_k among the full dataset, its mean μ_k and its covariance matrix Σ_k. Estimation of

1 Available at: https://scikit-learn.org/stable/modules/generated/sklearn.datasets.make_blobs.html.

parameters is classically done using the expectation maximization (EM) algorithm (Dempster et al. 1977).

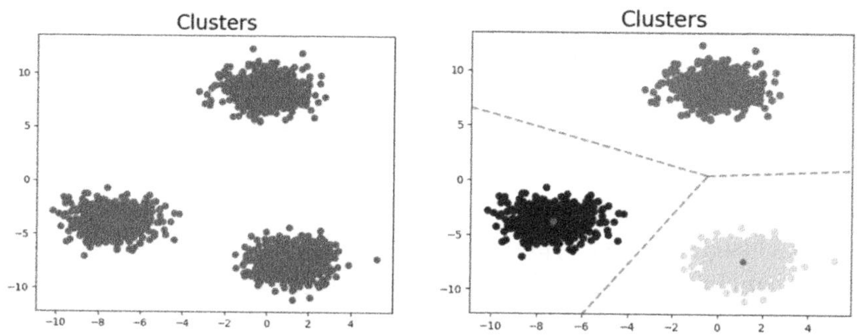

Figure 3.2. *Schematic representation of a K-means clustering. Initial unlabeled (same color) data points are shown in the left panel, on which the K-means is applied, choosing obviously $K = 3$ cluster. After convergence, the K-means model is represented on the right panel together with the blue points being the positions of the centroids. The red dashed lines highlights the separation in the three domains given the Euclidean distance metrics: the cluster space. Thus, with this model, each new point can now unambiguously be associated with one of the three clusters given its position in the feature space (here, the x- and y-axis in arbitrary units)*

The main issue of most clustering methods is the number of clusters that has to be set beforehand by a human. To circumvent the arbitrariness of choosing a number of cluster, and when there is a probability density function, the likelihood should be considered. It consists of measuring how close the samples are to the specified clustering distribution. However, this criterion will always be better for a higher number clusters, which lead to the classical problem of over-fitting. If the well-known elbow criterion was largely used to penalize such criterion, it is not related to any theoretical background and is always seen as very objective. Model selection criterion can be used, adapting the likelihood by penalizing it with respect to the dimension (related to the number of clusters) of each model. Usually, Akaike information criterion (AIC) (Akaike 1973) or Bayesian information criterion (BIC) (Schwarz 1978) are proposed, but the integrated completed likelihood (ICL) (Biernacki et al. 2000) should be preferred when focusing on clustering, because there is also a penalty term about the purity of each cluster based on a Shanon entropy term.

Density-based spatial clustering of applications with noise (DBSCAN) (Ester et al. 1996) is another famous clustering algorithm, known to be more robust. Observations lying in low-density regions are labeled as outliers. No assumptions on the form of the clusters is done, allowing for convex or non-convex clusters. Contrarily to K-means and GMM, DBSCAN does not directly rely on a given number of clusters, but on hyperparameters that are closely related to the number of clusters: thresholds on the similarity and outliers.

3.3.2. *Dimensionality reduction*

Dimensionality reduction techniques aim to project high-dimensional data into a lower dimensional space while preserving the underlying structure and relationships in the data at best. These techniques are widely used for visualization, data compression and noise reduction, for instance. This section focuses on the *principal component analysis* (PCA) and a nonlinear variant, namely, the *t-distributed stochastic neighbor embedding* (t-SNE).

PCA is a widely used linear technique that projects the data onto a lower dimensional subspace, maximizing the variance along the new axes. The principal components, which form the basis of the new subspace, are orthogonal and capture the directions of maximum variance in the data. The measure of the variance in the data is based on the calculation of the covariance matrix of the n data points in D-dimensional feature space. For two points x and y, it is given by:

$$\Sigma = \begin{pmatrix} \sigma_x^2 & \sigma_{xy} \\ \sigma_{yx} & \sigma_y^2 \end{pmatrix}, \tag{3.7}$$

where σ_x^2 and σ_y^2 are the variances of x and y, respectively, and σ_{xy} and σ_{yx} are the covariances between x and y. The variances read:

$$\sigma_x^2 = \frac{1}{D-1} \sum_{i=1}^{D} (x_i - \bar{x})^2, \tag{3.8}$$

where $\bar{x}$ is the mean of x, and the same for Y. The covariance between x and Y reads:

$$\sigma_{xy} = \sigma_{yx} = \frac{1}{D-1}\sum_{i=1}^{D}(x_i - \bar{x})(y_i - \bar{y}). \quad [3.9]$$

The lower dimensional representation $Y = \{y_1, y_2, \ldots, y_n\}$ with target dimensionality k of a dataset $X = \{x_1, x_2, \ldots, x_n\}$ of dimensionality D is obtained through algorithm 3.2. The n data points are projected on the space of the k eigenvectors whose eigenvalues are the largest and sorted by decreasing values.

Algorithm 3.2. Principal component analysis

Require: Data points of dimension D, $X = \{x_1, x_2, \ldots, x_n\}$ target dimensionality k

Ensure: Transformed data points $Y = \{y_1, y_2, \ldots, y_n\}$

Compute the mean vector $\mu = \frac{1}{n}\sum_{i=1}^{n} x_i$

Center the data points by subtracting the mean: $X_{centered} = X - \mu$

Compute the covariance matrix Σ

Calculate eigenvectors and eigenvalues of Σ: $(\lambda_1, v_1), \ldots, (\lambda_n, v_n)$

Sort eigenvectors by decreasing eigenvalues: $v_1, \ldots, v_n$

Select the top k eigenvectors: $V_k = [v_1, \ldots, v_k]$

Project the centered data onto the principal components: $Y = X_{centered}V_k$

Algorithm 3.2. illustrates the application of the PCA on a simple arbitrary bivariate random Gaussian dataset with 200 points using the `Scikit-learn` PYTHON package[2]. Interestingly, projection of the data on the first principal component (PC_1) is compared to a linear regression model learned on the same dataset. The good comparison indicates indeed that PC_1 carries the most important part of the information about the dataset.

2 Available at: https://scikit-learn.org/stable/modules/generated/sklearn.decomposition.PCA.html.

Contrarily to PCA, t-SNE is a nonlinear dimensionality reduction technique. It is often used for visualizing high-dimensional data, such as images or text embeddings, and is effective at revealing clusters and structures in the data, most often in two dimension and three dimension for visualization purposes (Van der Maaten and Hinton 2008). t-SNE works by minimizing the divergence between two probability distributions, one representing pairwise similarities in the high-dimensional space:

$$p_{j|i} = \frac{\exp(-||x_i - x_j||^2/2\sigma_i^2)}{\sum_{k \neq i} \exp(-||x_i - x_k||^2/2\sigma_i^2)} \quad [3.10]$$

$$p_{ij} = \frac{p_{j|i} + p_{i|j}}{2n} \quad [3.11]$$

using a Gaussian kernel and the other representing pairwise similarities in the low-dimensional space:

$$q_{ij} = \frac{(1 + ||y_i - y_j||^2)^{-1}}{\sum_{k \neq l}(1 + ||y_k - y_l||^2)^{-1}} \quad [3.12]$$

using the Student's t-statistics. The algorithm consists of minimizing the Kullback–Leibler divergence between the two similarity matrices using gradient descent (Hinton and Roweis 2002; Van der Maaten and Hinton 2008). The perplexity parameter p is a user-defined parameter that balances the focus on local and global structures in the data.

3.4. Application to particle dynamics

Unsupervised learning has a broad range of applications across various domains, as it can uncover hidden structures and relationships in data without the need for labeled examples. We restrict here to a specific case in materials science, namely, the atomic scale description of homogeneous nucleation, for which the unsupervised learning together with a topological descriptor was applied successfully for the first time very recently (Becker et al. 2022a, 2022b). Interestingly enough, this concerns particle dynamics and therefore could undoubtedly be applied at larger scales in geophysics.

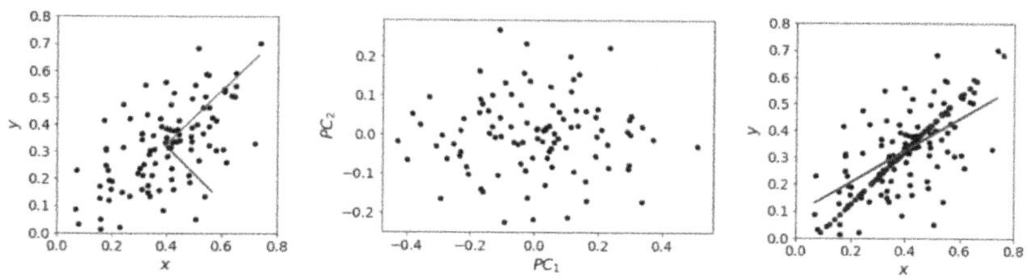

Figure 3.3. *Principal component analysis of simple bivariate random Gaussian dataset with two different variances, and having a linear correlation between the two random variables. The left panel shows the point cloud of the* 200 *point dataset in the* x-y *plane together with the eigenvector with the largest eigenvalue in red, and the second one in blue, from the PCA analysis. Center panel displays the dataset in the basis of these two eigenvectors, red and blue, respectively, denoted as* $PC1_1$ *and* PC_2*, and called the first and second principal components. The right panel shows in blue the PCA representation of the point cloud along* PC_1 *only. The red line is a linear regression model that was learned on the same dataset (see Chapter 2)*

Crystal nucleation, the early stages where the liquid-to-solid transition occurs upon undercooling, initiates at the atomic level on nanometer length and sub-picoseconds time scales and involves complex multidimensional mechanisms with local symmetry breaking that can hardly be observed experimentally in detail. In such cases, atomic-level simulations and more particularly MD with a suitable interaction model (Allen and Tildesley 2017) are the dedicated tools. However, reaching statistically meaningful nucleation events, large-scale simulations up to million to billion atom scale are required (Sosso et al. 2016).

In this section, an analysis is proposed for the MD simulations performed previously on the solidification of pure zirconium (Becker et al. 2020). The liquid state above the melting point (approximately $T_M = 2128$ K) was quenched down to the deep undercooled liquid at $T = 1250$ K, using a simulation box of 1 million atoms. The technical details of the MD simulation techniques are beyond the scope of the present chapter, and the reader is referred to (Becker et al. 2020). During the simulation, the liquid undergoes homogeneous nucleation on this isotherm as shown in Figure 3.4. In order to unveil their structural features during nucleation in such huge simulations, without a priori, an unsupervised learning approach founded on topological descriptors loaned from persistent homology concepts was built, and will be described in the following.

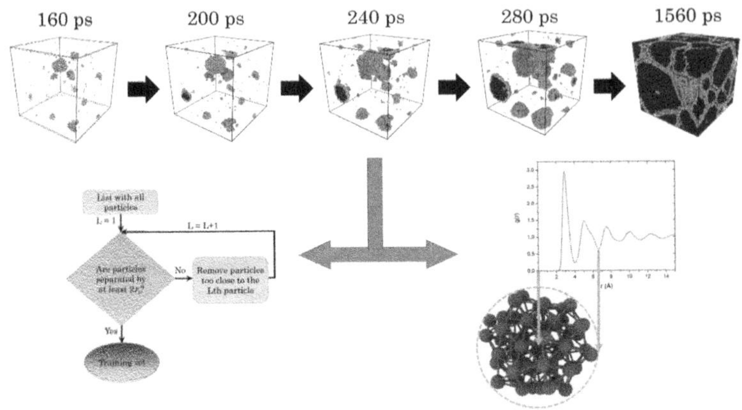

Figure 3.4. *Snapshot of a 1 million atom MD simulation of zirconium during nucleation along the $T = 1,250$ K isotherm in the undercooled states at different times after cooling (upper panels). Only atoms having a bcc or distorted bcc crystalline ordering are drawn as detected by the unsupervised topological learning approach (see text). From the snapshots, independent local atomic environment up to the second nearest neighbors defined by the second minimum of the pair-correlation function $g(r)$ (lower right panel), so-called local structures, are sampled by an the algorithm given in the lower left panel (Becker et al. 2022a)*

3.4.1. *Topological description of local structures*

Topological data analysis (TDA) (Chazal and Michel 2021; Motta 2018) is a growing mathematical field with applications in a wide range of other fields such as biology, computer science, physics and materials science. Persistent homology is an effective and flexible tool to study the underlying topological shapes of a point cloud. In atomic-scale simulations considered here, the point cloud usually corresponds to an atom assembly in a simulation box. Let us consider a dataset as the point cloud denoted $\mathcal{X}_0$ in a vector space and a parameter $r \geq 0$. The topological space $\mathcal{X}_r$ is defined by the union of all the balls (Euclidean or with another metric) of radius r, each of them being centered on each point of the point cloud. It consists of following the evolution of the topology of $\mathcal{X}_r$, or more precisely the persistence of its topological features as r grows from 0 about the initial set of points $\mathcal{X}_0$, to r big enough so that $\mathcal{X}_r$ has a the topology of a big ball containing all the points.

To encode or quantify the algebraic topology, a dedicated tool is homology (Hatcher 2002), which derives vector spaces $H_n(\mathcal{X})$, $n \geq 0$, from a space $\mathcal{X}$, generated by n-dimensional topological features. Zero-dimensional features correspond to connected components, one-dimensional ones correspond to "holes" in the space, two-dimensional features correspond to cavities and so on. Then persistent homology gives the persistence of these topological features, that is, their lifespan when r is growing. The tracking of this persistence is commonly represented either through a barcode, which is a graphical representation where each bar represents a topological feature, whose length indicates the lifespan (persistence) of that feature, or a persistence diagram (PD) where each topological feature is associated with the point (birth, death) in the plane as can be seen in Figure 3.5.

While persistent homology in molecular dynamics simulations usually consists of determining the PD from all atoms of the box as the point cloud, the originality here is to define from the PH a topological descriptor of the local environment of each atom, as shown in Figure 3.4. In this case, the point cloud is the local environment and contains a bit less than 100 atoms for the second neighbor shell. Finally, a PD is obtained for each local structure as a description of its topology, and it is associated with its central atom.

Among the various representations of these PDs as a vector, a classical method that has been successfully used to study three-dimensional shapes (Carrière et al. 2015) is chosen here. Each coordinate of the topological vector is associated with a pair of points (x, y) in a persistence diagram D for a fixed level of homology, except the infinite point, and is calculated by:

$$m_D(x,y) = \min\{\|x-y\|_\infty, d_\Delta(x), d_\Delta(y)\}, \tag{3.13}$$

where $d_\Delta(\cdot)$ denotes the ℓ^∞ distance to the diagonal, and those coordinates are sorted by decreasing order. The resulting topological space is high dimensional with often between 100 and 300 components.

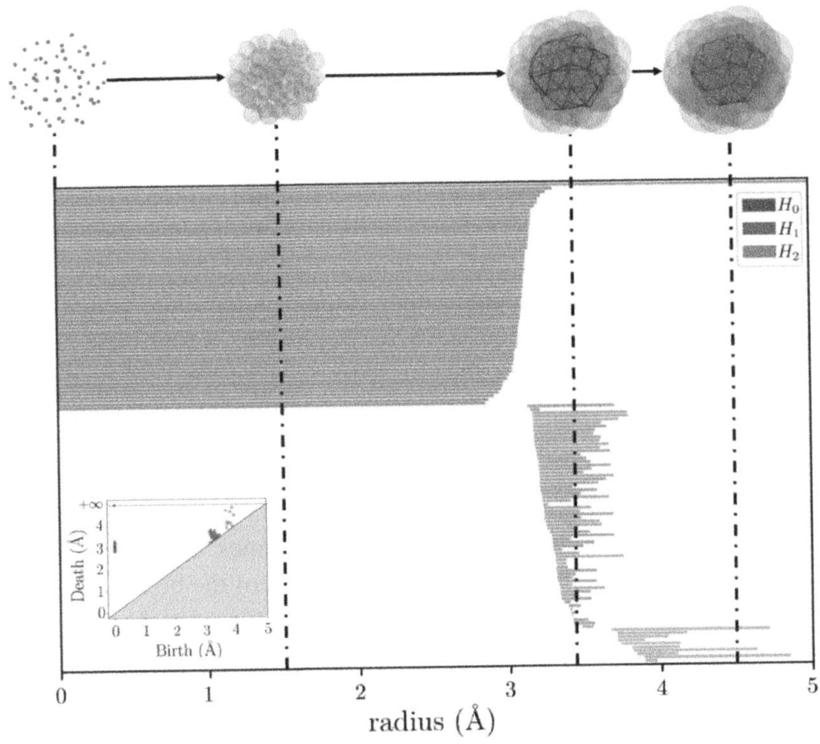

Figure 3.5. *Persistence of a local structure consisting of a central atom and its two neighborhood shells by means of a barcode description. The local structures with the ball growing as the radius increases are shown on the top. The barcode gives the lifespan of all topological features with colors corresponding to the homological dimension. In the bottom left corner, there is the corresponding persistent diagram (with the same corresponding colors for the dimensions)*

3.4.2. Clustering local environments during nucleation

For the clustering, a model-based method is used, namely, Gaussian mixture models (GMM) (Hastie et al. 2009) and its estimation by an expectation maximization (EM) algorithm (Dempster et al. 1977) as mentioned in section 3.3.1. The number of clusters is selected by ICL (Biernacki et al. 2000), a refinement for clustering of BIC (Schwarz 1978).

In a first step, a training set is built from a configuration during the nucleation, where the supercooled liquid coexists with crystalline nuclei, to capture all structural atomic events of interest. This is done by sampling approximately 5,000 atomic structures as described in Figure 3.4. Using PYTHON packages `gudhi` (Maria et al. 2014) and `ripser` (Tralie et al. 2018), the PD of each individual local atomic structures of the training set is computed up to homological dimensions H_0, H_1 and H_2 and the topological feature is calculated using equation [3.13].

In a second step, the unsupervised learning with the GMM is performed in the topological space iteratively with cluster numbers 2–30. The ICL criterion is computed for each of them giving rise to the curve shown in Figure 3.6. The optimal number of clusters is chosen from the minimum of ICL, leading to a model with seven clusters numbered C_1 to C_7 whose representative local structure (closest to the centroids) are shown in Figure 3.6. The inferred model from this method is called hereafter TDA-GMM.

Finally, with the learned TDA-GMM as such, each atom of the simulation box is assigned to C_1 to C_7 for any given configuration that enable one to identify and describe the structural properties of the system and morphological properties of the nuclei as can be seen on the specific configuration shown in Figure 3.6. Moreover, the crystal nucleation and its evolution as a function of time can be seen strikingly in Figure 3.4 where only the crystalline local structures (C_1 and C_2) are drawn. All the results and findings can be found in (Becker et al. 2022a, 2022b) and probably the main physical outcome in these works is that the liquid in the undercooled state appears very heterogeneous and the nucleation is triggered from fluctuations with lowest icosahedral ordering, always present in the liquid at various degrees, the latter being known as incompatible with the long-range ordering of the crystalline state (Frank 1952; Turnbull 1952).

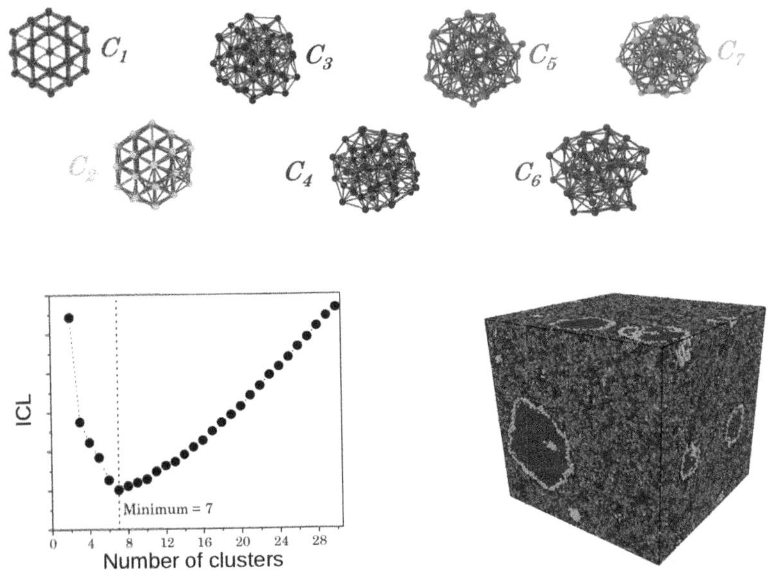

Figure 3.6. *TDA-GMM clustering model of undercooled liquid Zirconium during nucleation with seven clusters, C_1 to C_7, which are closest to the centroids. Evolution of the integrated completed likelihood (ICL) criterion (lower left panel) as a function of the number of clusters predicting the optimal number of seven clusters, trained on the configuration in the lower right panel. Adapted from (Becker et al. 2022b)*

To conclude this section, a question arises from the relevance of the TDA-GMM unsupervised modeling and especially the chosen number of clusters (seven for zirconium) in such a high-dimensional topological descriptor space with roughly $D \approx 200$. This can be visualized by performing a dimensionality reduction with the PCA or the t-SNE. Figure 3.7 displays a two-dimensional analysis for the two principal components of the topological vectors for all atoms in the simulation box shown in Figure 3.6. They are colored with the cluster they belong to. Interestingly, both methods lead to a similar representation, in which the clusterized local structures vary as a quasi continuum but do not overlap mainly (except very partially for C_4 and C_5) in the t-SNE representation. The PCA and tSNE distinguish the liquid on one component and the crystal on the second one. Clusters C_3, C_4 and C_5 that are known to retain partially liquid and crystalline orderings are located as expected at the crossover of the two components.

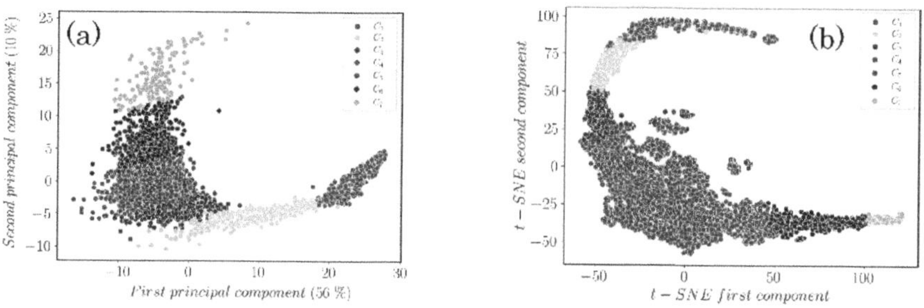

Figure 3.7. *Analysis of a configuration of undercooled Zr during nucleation with the PCA (left panel) and t-SNE (right panel) after each atom being labeled according to the TDA-GMM model with seven clusters*

3.5. Conclusion

This chapter was devoted to unsupervised learning approaches that emerged as a powerful approach for uncovering hidden patterns, relationships and structures in data without the need for labeled examples. The key concepts and techniques in unsupervised learning were discussed, focusing only on the basic approaches of clustering and dimensionality reduction, setting aside association rule mining and unsupervised deep learning methods, which could be treated as topics in themselves. The power of clustering was highlighted through a application in materials science that helped us to monitor efficiently the structural evolution during crystal nucleation of a liquid metal, and uncover hidden correlations without a priori in this process from the huge amount of data of these large-scale molecular dynamics simulations with millions of atoms. Such an unsupervised approach is deemed to be sufficiently general to be transposed to discrete particle dynamics at a larger scale in geophysics.

3.6. Acknowledgements

We acknowledge the CINES and IDRIS under Project No. INP2227/72914, as well as CIMENT/GRICAD for computational resources. This work was performed within the framework of the Centre of Excellence of Multifunctional Architectured Materials CEMAM-ANR-10-LABX-44-01 funded by the "Investments for the Future" Program. This work has been partially supported by MIAI@Grenoble Alpes (ANR-19-P3IA-0003). Discussions within the French collaborative network in high-temperature thermodynamics GDR CNRS3584 (TherMatHT) and in artificial intelligence in materials science GDR CNRS 2123 (IAMAT) are also acknowledged.

3.7. References

Akaike, H. (1973). *Information Theory and an Extension of the Maximum Likelihood Principle*. Springer, New York.

Allen, M.P. and Tildesley, D.J. (2017). *Computer Simulation of Liquids*. Oxford University Press, Oxford.

Becker, S., Devijver, E., Molinier, R., Jakse, N. (2020). Glass-forming ability of elemental zirconium. *Physical Review B*, 102, 104205.

Becker, S., Devijver, E., Molinier, R., Jakse, N. (2022a). Unsupervised topological learning for identification of atomic structures. *Physical Review*, 105, 045304.

Becker, S., Devijver, E., Molinier, R., Jakse, N. (2022b). Unsupervised topological learning approach of crystal nucleation. *Scientific Reports*, 12, 1–9. doi: 10.1038/s41598-022-06963-5.

Biernacki, C., Celeux, G., Govaert, G. (2000). Assessing a mixture model for clustering with the integrated completed likelihood. *IEEE Transactions on Pattern Analysis and Machine Intelligence*, 22(7), 719–725.

Carlsson, G. (2020). Topological methods for data modelling. *Nature Reviews Physics*, 2, 697–708. doi: 10.1038/s42254-020-00249-3.

Carrière, M., Oudot, S.Y., Ovsjanikov, M. (2015). Stable topological signatures for points on 3D shapes. *Eurographics Symposium on Geometry Processing*, 34, 1–12.

Ceriotti, M. (2019). Unsupervised machine learning in atomistic simulations, between predictions and understanding. *The Journal of Chemical Physics*, 150(15), 150901.

Chazal, F. and Michel, B. (2021). An introduction to topological data analysis: Fundamental and practical aspects for data scientists. *Frontiers in Artificial Intelligence*, 4 [Online]. Available at: https://www.frontiersin.org/articles/10.3389/frai.2021.667963.

Dempster, A.P., Laird, N.M., Rubin, D.B. (1977). Maximum likelihood from incomplete data via the EM algorithm. *Journal of the Royal Statistical Society: Series B (Methodological)*, 39(1), 1–22.

Efron, B. and Hastie, T. (2021). *Computer Age Statistical Inference, Student Edition: Algorithms, Evidence, and Data Science*, vol. 6. Cambridge University Press, Cambridge.

Ester, M., Kriegel, H.-P., Sander, J., Xu, X. (1996). A density-based algorithm for discovering clusters in large spatial databases with noise. In *Proc. of 2nd International Conference on Knowledge Discovery and Data Mining*, AAAI Press, Portland, 226–231.

Frank, F.C. (1952). Supercooling of liquids. *Proceedings of the Royal Society of London. Series A. Mathematical and Physical Sciences*, 215(1120), 43–46.

Hastie, T., Tibshirani, R., Friedman, J.H. (2009). *The Elements of Statistical Learning: Data Mining, Inference, and Prediction*, vol. 2. Springer, New York.

Hatcher, A. (2002). *Algebraic Topology*. Cambridge University Press, Cambridge.

Hey, T. and Trefethen, A. (2020). The fourth paradigm 10 years on. *Informatik Spektrum*, 42, 441–447.

Hey, T., Tansley, S., Tolle, K., Gray, J. (2009). *The Fourth Paradigm: Data-Intensive Scientific Discovery*. Microsoft Research, Redmond, WA [Online]. Available at: https://www.microsoft.com/en-us/research/publication/fourth-paradigm-data-intensive-scientific-discovery/.

Hinton, G.E. and Roweis, S. (2002). Stochastic neighbor embedding. *Advances in Neural Information Processing Systems*. MIT Press, Cambridge, 15, 833–840.

Huang, Y., Wu, Z., Wang, L., Tan, T. (2013). Feature coding in image classification: A comprehensive study. *IEEE Transactions on Pattern Analysis and Machine Intelligence*, 36(3), 493–506.

Karapiperis, K. and Andrade, J. (2021). Nonlocality in granular complex networks: Linking topology, kinematics and forces. *Extreme Mechanics Letters*, 42, 101041 [Online]. Available at: https://www.sciencedirect.com/science/article/pii/S2352431620302455.

Khalid, S., Khalil, T., Nasreen, S. (2014). A survey of feature selection and feature extraction techniques in machine learning. In *2014 Science and Information Conference*. IEEE.

Maria, C., Boissonnat, J.-D., Glisse, M., Yvinec, M. (2014). The GUDHI library: Simplicial complexes and persistent homology. Research Report RR-8548, INRIA [Online]. Available at: https://inria.hal.science/hal-01005601v2/document.

Motta, F.C. (2018). Topological data analysis: Developments and applications. *Advances in Nonlinear Geosciences*, 6, 369–391.

Nothman, J., Qin, H., Yurchak, R. (2018). Stop word lists in free open-source software packages. In *Proceedings of Workshop for NLP Open Source Software (NLP-OSS)* [Online]. Available at: https://nlposs.github.io/.

Pirashvili, M., Steinberg, L., Guillamon, F.B., Niranjan, M., Frey, J.G., Brodzki, J. (2018). Improved understanding of aqueous solubility modeling through topological data analysis. *Journal of Cheminformatics*, 10, 1–14. doi: 10.1186/s13321-018-0308-5.

Schmidt, J., Marques, M.R., Botti, S., Marques, M.A. (2019). Recent advances and applications of machine learning in solid-state materials science. *npj Computational Materials*, 5(1), 83.

Schwarz, G. (1978). Estimating the dimension of a model. *The Annals of Statistics*, 6(2), 461–464. doi: 10.1214%2Faos%2F1176344136.

Sosso, G.C., Chen, J., Cox, S.J., Fitzner, M., Pedevilla, P., Zen, A., Michaelides, A. (2016). Crystal nucleation in liquids: Open questions and future challenges in molecular dynamics simulations. *Chemical Reviews*, 116, 7078–7116.

Tralie, C., Saul, N., Bar-On, R. (2018). Ripser.py: A lean persistent homology library for Python. *Journal of Open Source Software*, 3(29), 925. doi: 10.21105/joss.00925.

Turnbull, D. (1952). Kinetics of solidification of supercooled liquid mercury droplets. *The Journal of Chemical Physics*, 20(3), 411–424.

Van der Maaten, L. and Hinton, G. (2008). Visualizing data using t-SNE. *Journal of Machine Learning Research*, 9(11), 2579–2605.

Walker, D.M. and Tordesillas, A. (2010). Topological evolution in dense granular materials: A complex networks perspective. *International Journal of Solids and Structures*, 47(5), 624–639 [Online]. Available at: https://www.sciencedirect.com/science/article/pii/S0020768309004272.

4

Classification Techniques in Machine Learning

Noel Jakse
SIMaP, CNRS, Grenoble INP, Université Grenoble Alpes, France

4.1. Introduction

At the turn of the 21st century, machine learning (ML) techniques aroused a growing interest across various scientific disciplines due to their inherent capability to identify trends, forecast outcomes and draw invaluable knowledge from vast and intricate datasets (Hey and Trefethen 2020; Hey et al. 2009). Tools designed for collecting data online, scientific apparatus and computer-simulated environments are generating data at an exponential rate. This has spurred new scientific methodologies for dissecting and structuring enormous volumes of data, while simultaneously detecting subtle effects that were previously hidden. These techniques have become increasingly

For a color version of all the figures in this chapter, see www.iste.co.uk/stefanou/machine1.zip.

Machine Learning in Geomechanics 1,

prevalent in areas such as materials science (Schmidt et al. 2019) and geophysics (Karapiperis and Andrade 2021).

ML approaches can be primarily grouped into three categories: supervised learning, unsupervised learning and semi-supervised learning. Each category presents unique ways of deriving knowledge from data, each possessing their own benefits and limitations. Among these, supervised learning is widely adopted, where a model is trained using labeled data, consisting of both input attributes and associated output labels. The model is trained to associate inputs with their corresponding outputs, enabling it to make predictions for novel, unobserved data.

This chapter is devoted to classification, which has numerous applications across many domains, including spam detection, medical diagnosis and image recognition to name only some of the most emblematic ones. Like regression, being the subject of Chapter 2, classification belongs to supervised learning techniques. The objective is to build models that can predict the class label of unseen data, after being trained on a representative set of labeled data. An application of classification of crystalline structure in materials science and more particularly soft matter is briefly mentioned to highlight the power of the method (Reinhart et al. 2017; Reinhart and Panagiotopoulos 2018).

ML algorithms generally require large training sets to get high performance, but labeling such large amounts of data represents substantial human effort and often costly experimental or numerical resources. Active learning (AL) proposes effective ways to reduce the size of training sets while maintaining the same performance. It represents an attractive strategy in scenarios where data labeling is expensive or time consuming, which is particularly the case for materials exploration in the vast composition space to accelerate their discovery and development (Lookman et al. 2019).

This chapter does not deal with neural networks and deep learning techniques for classification. They gained significant momentum in recent years due to their ability to learn complex hierarchical representations from large datasets and take advantage of the tremendous increase in computer power.

The remaining part provides an overview of the most commonly used classification techniques, their strengths and limitations, as well as important technical aspects like pre-processing, evaluation metrics and cross-validation. More advanced techniques for classification are then presented briefly, such as ensemble methods. Finally, the basic aspects of AL is presented.

4.2. Classification techniques

4.2.1. *General considerations*

In this section, some of the most popular classification techniques and their characteristics will be reviewed. The general classification problem can be described as follows. Let us consider a set of n labeled data:

$$\{X_i, y_i, i = 1, \ldots, n\}. \qquad [4.1]$$

Each data instance X_i is labeled with a class y_i. Without losing generality, let us consider that X_i for all i are points in the d-dimensional real space $\mathbb{R}^d$. For classification, labels take discrete values and two situations are considered:

– binary classification with $y_i \in \mathcal{C} = \{-1, 1\}$, thus having a so-called *positive class* and *negative class*. Note that alternately classes can be defined as $y_i \in \mathcal{C} = \{0, 1\}$;

– multi-class classification with K classes when $y_i \in \mathcal{C} = \{1, \ldots, K\}$.

It is worth mentioning that if $y_i \in \mathbb{R}$ reduces to the case of regression, which will be the subject of Chapter 2, and needs a model such as in neural networks (see Chapters 2 and 3 of volume 2). The objective of classification is to predict the class value $y_+ \in \mathcal{C}$ for each new X_+ not in the training set. For this purpose, the training data are used to construct a classifier $\hat{c}$ such that:

$$y_+ = \hat{c}(X_+) \tag{4.2}$$

One of the main aspects of classification is that it relies on labeled data to learn the relationship between input data and class labels that are identified in the training set. This is in contrast with unsupervised learning that deals with unlabeled data and is treated in Chapter 3.

Therefore, labeling of data plays a crucial role in supervised learning either for classification or regression, as they provide to the ML algorithm the relationships between inputs and output labels. The quality and quantity of labeled data can significantly impact the performance of the obtained model, so-called the *classifier*. This is why acquiring as large, diverse and representative a dataset as possible is often essential.

Finally, classification methods imply categorizing data instances into distinct classes. There are subjective aspects influencing the model, which stem from human decisions, interpretations and biases in the choice of the classes, as illustrated in Figure 4.1. The way a classification problem is formulated and how classes are defined such as animals, their color and type often depends on the domain knowledge, goals and preferences of the researchers. It leads to discrepancies between different formulations of the same problem.

4.2.2. *Typical workflow for classification*

The general ML scheme is described in Chapter 3, devoted to unsupervised learning. In Figure 4.2, the typical workflow for classification is described schematically. After having clearly defined the problem that needs to be solved including the choice of the labeling, the first crucial step consists of data collection, preprocessing to prepare the training and test data that will be used to train and evaluate the ML model. Collection of raw data can be done from various sources, such as databases, APIs, sensors or web scraping.

Then preprocessing typically includes data cleaning (e.g. handling missing values, removing duplicates or fixing errors). The second step consists of the data transformation such as normalization, standardization or feature transformation and extraction as well as associated dimensionality reduction (see Chapter 3). In particular, normalization typically scales features to a range of $[0, 1]$, while standardization scales features to have a mean of 0 and a standard deviation of 1.

Figure 4.1. *Classification can be subjective. Image generated by the Author with OpenAI DALLE-E with order: realistic painting with white and black horses, pigs, cows and chicken in a field*

With the data prepared as such, the third step is to select an appropriate ML algorithm and train the model that best suits the problem formulation and the characteristics of the data. Finally, after training the model, it is important to evaluate its performance on a separate dataset (test set) that was not used during training. This helps assess their ability to generalize new, unseen data. Various evaluation metrics, such as *accuracy*, *precision*, *recall* and *F1-score*, can be used to quantify the classification model's performance, depending on the type of ML task as will be seen in section 4.2.3. In addition to evaluating models' performance, it might be necessary to tune the parameters of the learning algorithms, the so-called *Hyperparameter*. Tuning them involves searching for the optimal combination of their values that yields the best model performance. This typically consists of using techniques such as grid search, random search or Bayesian optimization (Rasmussen and Williams 2006; Shields et al. 2021).

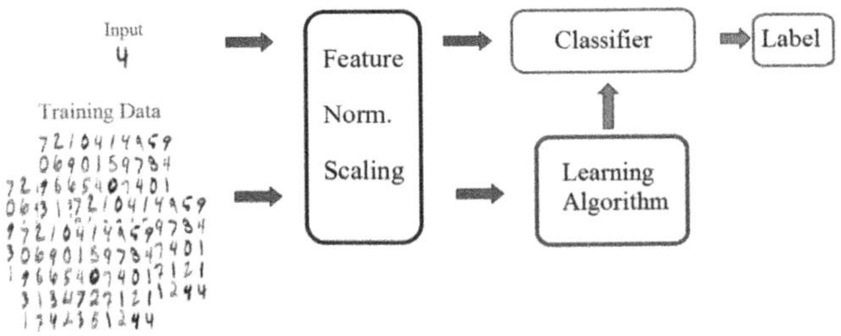

Figure 4.2. *Schematic representation of a typical workflow for classification. A training dataset is build, featurized, scaled and normalized. A classifier is then trained with a chosen learning algorithm. It can then be used to recognize unseen handwritten digits. Note that the new digits shown to the classifier for recognition also have to be transformed with the same feature model, scaling and norm used for the training. Example: https://archive.ics.uci.edu/ml/datasets/Optical+Recognition+of+Handwritten+Digits*

4.2.3. *Evaluation metrics*

4.2.3.1. *Standard metrics*

Accuracy is the most straightforward and simple evaluation metric, measuring the proportion of correctly classified instances in the dataset. It is defined as the ratio:

$$\text{Accuracy} = \frac{\text{Number of correct predictions}}{\text{Total number of predictions}}. \qquad [4.3]$$

However, accuracy can be misleading in cases of class imbalance, as it may overemphasize the performance on the majority class. A confusion matrix for a binary classification problem can then be considered. The confusion matrix is a table that summarizes the performance of the classifier by showing the counts of true positive (TP), true negative (TN), false positive (FP) and false negative (FN). A TP is a case in which the classifier correctly predicts the positive class, while FP is a case in which the classifier incorrectly predicts the positive class. The same definition holds for the TN and FN. The accuracy can then be calculated as:

$$\text{Accuracy} = \frac{\text{TP} + \text{TN}}{\text{TP} + \text{TN} + \text{FP} + \text{FN}}. \qquad [4.4]$$

Moreover, this matrix allows for a deeper understanding of the model's strengths and weaknesses of the chosen classes in the model. To do so, *precision* and *recall* metrics can be used to account for FPs and FNs, respectively. Precision, also known as positive predictive value, is the ratio of TP predictions to the total number of positive predictions made. Precision can be calculated as:

$$\text{Precision} = \frac{\text{TP}}{\text{TP} + \text{FP}}. \qquad [4.5]$$

Recall, also known as *sensitivity* or *TP rate*, is the ratio of TP predictions to the total number of actual positive instances. Recall can be calculated as:

$$\text{Recall} = \frac{\text{TP}}{\text{TP} + \text{FN}} \qquad [4.6]$$

Finally, the *F1-score* is the harmonic mean of precision and recall, providing a balanced measure of both metrics. It can be calculated as:

$$\text{F1-score} = 2 \times \frac{\text{Precision} \times \text{Recall}}{\text{Precision} + \text{Recall}}. \qquad [4.7]$$

The F1-score ranges from 0 to 1, with 1 being the best possible score, indicating perfect precision and recall.

The area under the ROC curve (AUC-ROC) is an evaluation metric often used for binary classification problems that at least should be mentioned. It measures the performance of a classifier across different decision thresholds, plotting the TP rate (recall) against the FP rate. AUC-ROC provides a single scalar value representing the overall performance of the classifier, with a value of 1 indicating perfect classification and a value of 0.5 corresponding to random chance.

4.2.3.2. *Cross-validation*

Cross-validation or *folding* is a concept used for evaluating the performance of ML models in general. It consists of estimating statistically the average of an evaluation metric of the model, such as accuracy, over multiple training and testing sets. It consists of dividing the dataset into multiple subsets and iteratively training and testing the model on different combinations of these subsets so-called *folds*. This process allows for a more reliable estimation of the model's performance, especially when the amount of available data is limited. One advantage is that it reduces the likelihood of going into *overfitting* and provides a better understanding of the model's generalization capabilities.

Algorithm 4.1. Cross-validation

Require: Randomly shuffle the dataset.
Ensure: Divide the dataset into k equally sized (or approximately equal) folds.
for $i = 1, 2, ..., k$ **do**
Use fold i as the testing set, and the remaining k-1 folds as the training set.
Train the model on the training set.
Evaluate the model on the testing set, and compute the evaluation metric.
end for
Calculate the average of the evaluation metric values over all k iterations.

In most of the cases, the k-fold cross-validation approach is used, for which the dataset is randomly divided into k equally sized subsets, thus leading to k train and test steps as described in Algorithm 4.1.

Each step uses a different fold as the testing set and the remaining $k-1$ folds as the training set. The evaluation metric (e.g. accuracy, precision, recall or F1-score mentioned just above) is computed for each iteration, and the final result is obtained by averaging the metric values over all iterations. Typically, k is set to values such as 5 or 10, as these values were empirically shown to provide a good balance between bias and variance.

Leave-one-out cross-validation (LOOCV) is a limiting case of k-fold cross-validation, where k is set equal to the number of instances in the dataset. In other words, each instance is used as a testing set exactly once. Therefore, LOOCV can be computationally expensive, especially for large datasets, as it requires training and evaluating the model as many times as there are instances in the dataset.

4.2.4. *Standard classification algorithms*

4.2.4.1. *k-Nearest neighbors*

Let us first consider the k-nearest neighbors (k-NN), which is a non-parametric, instance-based classification algorithm, and is among the simplest of all ML algorithms, both conceptually and in terms of implementation. Given a query point x in a training set $\mathcal{T}$ with m samples, $\{(x^{(i)}, y^{(i)}), i = 1, \ldots, m\}$, each with a label y, the objective of k-NN is to assign x the label most commonly associated with its k-NNs in $\mathcal{T}$.

Algorithm 4.2 consists of computing the distance $d(x, x^{(i)})$ between x and each training instance $x^{(i)}$, $i = 1, \ldots, m$. The choice of distance function typically depends on the problem under consideration. Common choices include the Euclidean distance and the Manhattan distance as described in Chapter 3. The k instances $x^{(i_1)}, \ldots, \boldsymbol{x}^{(i_k)}$ that are closest to x are identified and the most common label among these k neighbors is assigned to x . The assigned label $\hat{y}$ can be formally expressed as:

$$\hat{y} = \arg\max_{y} \sum_{j=1}^{k} I(y = y^{(i_j)}), \qquad [4.8]$$

where $I(\cdot)$ is the indicator function, which equals 1 if the condition inside the parenthesis is true and 0 otherwise.

Algorithm 4.2. k-Nearest neighbors

Require: Training set $\mathcal{T}$, query point $\boldsymbol{x}$, number of neighbors k
  for each instance $\boldsymbol{x}^{(i)}$ in $\mathcal{T}$ **do**
    Compute $d(\boldsymbol{x}, \boldsymbol{x}^{(i)})$
  end for
  Identify the k instances $\boldsymbol{x}^{(i_1)}, \ldots, \boldsymbol{x}^{(i_k)}$ with smallest distances to $\boldsymbol{x}$
  $\hat{y} = \arg\max_y \sum_{j=1}^{k} I(y = y^{(i_j)})$ **return** $\hat{y}$

The choice of *hyper-parameter* k is critical to the performance of the algorithm. A small value of k means that noise will have a higher influence on the result and a large value makes it computationally expensive. Usually, an odd number is chosen if the number of classes is 2 (binary classification) and another simple approach to select k is set $k = \sqrt{n}$.

As an example, Figure 4.3 shows the results on a binary classification task using the k-NN classifier. Let us consider a dataset consisting of $n = 600$ points distributed in a plane with axis x_1 and x_2 using a Gaussian bivariate distribution of similar variance as shown in Figure 4.3(a). It can be produced using the following code:

```
def dataset_fixed_cov(n=300,dim=2):
    np.random.seed(0)
    # centers
    C = np.array([[0.0, -0.23], [0.83, 0.23]])
    X = np.r_[
        np.dot(np.random.randn(n, dim), C),
        np.dot(np.random.randn(n, dim), C) + np.array([1, 1]),]
    y = np.hstack((np.zeros(n), np.ones(n)))
    return X, y

N_sample = 300
N_dim =2
X,y = dataset_fixed_cov(N_sample,N_dim)
```

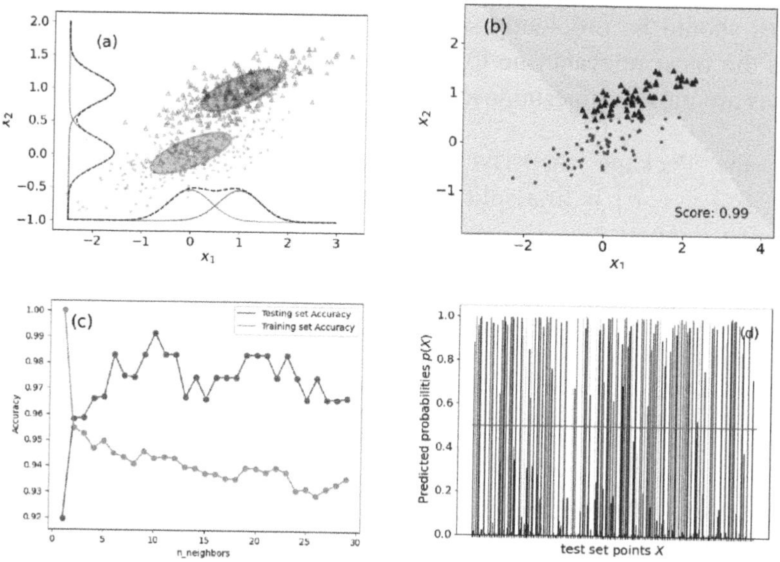

Figure 4.3. *Classification task on a Gaussian bivariate distribution of $n = 600$ points with the same variance in two dimensions x_1 and x_2, using the k-NN classifier. (a) The dataset with labels in red and blue, respectively, presents the positive and negative classes. Yellow stars are the respective centroids and ellipses materialize the standard deviation of the Gaussian distributions about the centroids. Red and blue lines are the Gaussian distributions of the respective classes projected onto x_1 and x_2 using a linear discriminant analysis. The dashed line is the total distribution. (b) Test set with positive and negative classes in red and blue, respectively, as well as predictions of the best k-NN classifier, namely, $k = 10$, highlighted by the light red and blue background. (c) F1-score of the test set and training set for k-NN classifiers, trained with k in the range $[1, 30]$. (d) Predicted probabilities of each points in the test set and the horizontal line represent the decision line with probability $p = 0.5$ used for attributing class labels on the unseen points*

Positive and negative class labels are colored in red and blue, respectively. In a first step, the dataset is plotted as a point cloud and analyzed with linear discriminant analysis (LDA, similar to the principal component analysis described in Chapter 3). It shows indeed that the points of each class follow a Gaussian distribution along x_1 as well as x_2. The ellipses, with radii corresponding to the distance of the standard deviation from the centroids, do not overlap, indicating that the classes should be efficiently separated. The LDA further reveals that x_2 is the most relevant one to discriminate the classes, while the two classes are mostly superimposed along x_1.

Using the Package SCIKIT-LEARN[1] in PYTHON, the dataset $\{X_i, y_i, i = 1, \ldots, n\}$ is first split into a training set and a test set as the following snippet:

```
from sklearn.model_selection import train_test_split

random_state = 42
X_train, X_test, y_train, y_test = train_test_split(X,y,
        test_size=0.20, random_state=random_state)
```

The usual size of the training set is 80% of the data points, leaving 20% of them for the test set to measure the performance of the trained model. Here, the random state is fixed in order to compare different models on strictly the same training set. The k-NN model is set up with desired hyper-parameters, leaving most of them to the default values, and fixing only $k = 10$ neighbors corresponding to the best classifier shown in Figure 4.3(b).

```
from sklearn.neighbors import KNeighborsClassifier

n_neighbors = 10
knn = KNeighborsClassifier(n_neighbors=n_neighbors)
```

The model is trained by providing to the learning algorithm inputs X and outputs y of the training set.

```
knn.fit(X_train, y_train)
```

Once trained, the model can be used to predict the class of unseen data; let us say those of the test set at first from the command.

1 Available at: https://scikit-learn.org/.

```
y_pred = knn.predict(X_test)
```

The performance of the model is measured by comparing the predictions on the text set against the true known values, using the F1-score for instance:

```
from sklearn.metrics import f1_score

Score = f1_score(y_test, y_pred)
```

Here, Figure 4.3(b) indicates the value of 0.99, which is close to the perfect value of 1. Prediction made now on the whole (x_1, x_2) plane leads to the colored background corresponding to the obtained classes. The boundary separating the two colored areas is the decision line, which separates appropriately the two classes horizontally in the vicinity of the dataset as inferred initially from the LDA.

The model with $k = 10$ was not chosen by chance. A simple *model selection* was performed by iterating on k-values between 1 and 30, training the k-NN for each k, and evaluating the performance using the F1-score. Figure 4.3(c) displays the performance of k-NN as a function of k measured on the test set (blue curve), showing that $k = 10$ performs the best in that case. The general rule $k = \sqrt{n}$ with $n = 240$ (the size of the training set) gives a value of 14, which is a bit higher than the best value of k but remains relevant. A *validation test* can be performed (orange curve) by predicting the classes on the training set itself, which can be indicative on how the model learns. It can be seen that the validation score becomes somewhat stationary above $k = 15$. An increase toward high values is often characteristic to over-fitting, which is typically the case for $k = 1$. The model learned the training data very well but when it comes to the prediction, it appears to perform the worst.

Finally, Figure 4.3(d) describes the mechanism by which the classes are chosen by the algorithm, given the predicted probability obtained as:

```
p = knn.predict_proba(X_test)
```

For each point to predict, the negative class is chosen if $p(X) > 0.5$, and positive class is chosen otherwise. Figure 4.3(d) shows that most of the points are classified correctly.

4.2.4.2. *Logistic and softmax regressions*

Logistic regression is a linear model for *binary classification* that uses the logistic function to model the probability of an instance belonging to a particular class. The model is trained by minimizing the negative log-likelihood (cross-entropy loss) using gradient descent or Newton–Raphson method. For prediction, the model computes the probability of the instance belonging to the positive class and assigns the class label based on a threshold value, typically 0.5.

The method uses the logistic function, also known as the sigmoid function denoted as $\sigma(z)$:

$$\sigma(z) = \frac{1}{1 + e^{-z}}. \qquad [4.9]$$

The function takes a real-valued input z and maps it to a value between 0 and 1, making it suitable for modeling probabilities. Given a set X of input features $\{x_1, x_2, ..., x_n\}$, the logistic regression model computes the probability $P(Y = 1|X)$ of an instance belonging to the positive class in Y using the following equation:

$$P(Y = 1|X) = \sigma(w_0 + w_1 x_1 + w_2 x_2 + ... + w_n x_n), \qquad [4.10]$$

where $w_0, w_1, \ldots, w_n$ are the model parameters (also known as *weights*) that need to be estimated from the training data, and σ denotes the logistic function given by equation [4.9].

Optimization of the parameters is based on the maximum likelihood estimation (MLE) method. The objective is to find the parameter values that maximize the likelihood of the observed data. For a given dataset with m samples, $\{(x^{(i)}, y^{(i)}), i = 1, \ldots, m\}$, the likelihood function is defined as:

$$L(w) = \prod_{i=1}^{m} P(Y = y^{(i)} | X = x^{(i)}). \quad [4.11]$$

To simplify the optimization problem using a gradient-based method, the negative log-likelihood is minimized instead:

$$J(w) = -\sum_{i=1}^{m} [y^{(i)} \log(\hat{y}^{(i)}) + (1 - y^{(i)}) \log(1 - \hat{y}^{(i)})], \quad [4.12]$$

where $\hat{y}^{(i)}$ denotes the predicted probability of the positive class for instance i. The corresponding algorithm is described in Algorithm 4.3.

Algorithm 4.3. Logistic regression

Require: Initialize the model parameters $w_0, w_1, ..., w_n$ with small random values or zeros

repeat

Calculate the predicted probability of the positive class for each instance using the logistic function.

Compute the negative log-likelihood and its gradient with respect to the model parameters.

Update the model parameters using the gradient descent update rule.

until convergence or the maximum number of iterations is reached.

To classify an unseen instance, its predicted probability of belonging to the positive class is compared to a threshold value, typically 0.5 as mentioned above. If the probability is greater than or equal to the threshold, the instance is classified as positive; otherwise, it is classified as negative, as:

$$\hat{c} = \begin{cases} 1 & \text{if } P(Y = 1|X) \geq 0.5 \\ 0 & \text{otherwise} \end{cases} \quad [4.13]$$

For multi-class problems, the classifier reads:

$$\hat{c} = \text{argmax}_{k\in\{1,\ldots,K\}}\#\{i : Y_i = k\} \qquad [4.14]$$

The limitations of this method include sensitivity to outliers, inability to model nonlinear relationships, and the assumption of class label independence for multi-class problems.

As mentioned above, the extension of logistic regression to handle multi-class classification problems directly is the so-called *softmax regression*. In this scheme, the probability of an instance belonging to each class is estimated using a generalized logistic function for multiple classes. The probability of the instance belonging to class k is calculated using:

$$P(Y = k|X) = \frac{e^{(w_k^T x)}}{\sum_{j=1}^{K} e^{(w_j^T x)}}, \qquad [4.15]$$

where w_k^T represents the weight vector for class k.

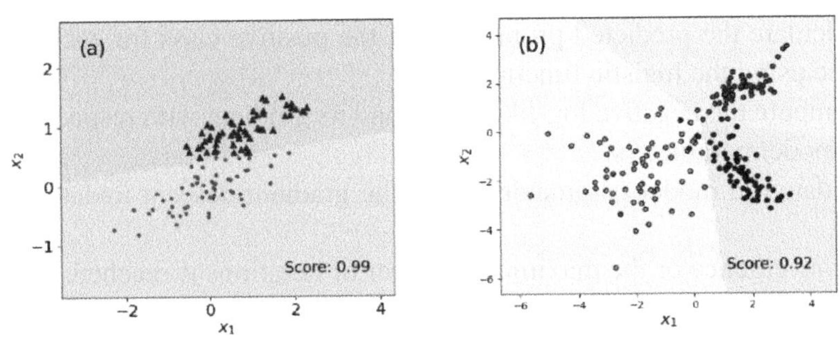

Figure 4.4. *Classification tasks using the logistic regression classifier. (a) Binary classification with the same dataset and same coloring as in Figure 4.3. (b) Multi-class classification using the softmax method with a dataset containing three classes in the two-dimensional space colored in red, blue and green, respectively, for classes 1, 2 and 3. In both cases, boundary decisions are materialized by the background colors corresponding to the predictions*

Figure 4.4(a) shows the logistic regression for the binary classification using the same dataset as for the k-NN models. The same scheme as described before is applied here (not repeated). The logistic regression model is included in the SCIKIT-LEARN and setup as the following:

```
from sklearn.linear_model import LogisticRegression

logreg = LogisticRegression(solver='liblinear',
    multi_class='ovr')
```

The decision curve is linear, as expected by the required solver, and separates the two classes horizontally as inferred by the LDA with a score of 0.99. For hyperparameter "`ovr`", see below.

Figure 4.4(b) shows a multi-class logistic regression with a dataset created this time with the very handy `make_classification` of SCIKIT-LEARN with three classes using:

```
from sklearn.datasets import make_classification

n_classes = 3
Xm, ym = make_classification(n_samples=1000,
                             n_features=2,
                             n_redundant=0,
                             n_classes= n_classes,
                             n_clusters_per_class = 1,
                             class_sep = 1.7)
```

The reader is referred to the documentation for the parameters of the `make_classification` function. The set up of the model is done with the same function as for the binary classification, but with different hyper-parameters.

```
from sklearn.linear_model import LogisticRegression

logreg = LogisticRegression(multi_class='multinomial',
    solver='lbfgs')
```

Here, the (x_1, x_2) plane is divided into three areas by linear decision boundaries and a score of 0.92.

4.2.4.3. *Support vector machines*

Support vector machines (SVMs) represent a class of linear models that aim to find the optimal separating hyperplane between classes by maximizing the margin between support vectors, that is, those instances closest to the decision boundary. They are particularly well suited for high-dimensional and small-to-medium-sized datasets. However, for large datasets, SVMs can be computationally expensive.

Let us consider a training set $\{(\boldsymbol{x}^{(i)}, y^{(i)}), i = 1, \ldots, m\}$, where $\boldsymbol{x}^{(i)} \in \mathbb{R}^n$ represents the feature vector of instance i, and $y^{(i)} \in \{-1, 1\}$ is its corresponding class label in binary classification. The goal of SVM is to find the optimal separating hyperplane that maximizes the *margin* between the classes. The equation of the hyperplane can be represented as:

$$\boldsymbol{w}^T \boldsymbol{x} + b = 0, \qquad [4.16]$$

where $\boldsymbol{w} \in \mathbb{R}^n$ is the weight vector normal to the hyperplane and b is the bias. The *margin* is defined as the distance between the hyperplane and the closest instances from each class, the so-called *support vectors*. The margin M can be written as:

$$M = \frac{2}{|\boldsymbol{w}|}. \qquad [4.17]$$

Finding the optimal hyperplane given by equation [4.16] consists of solving the following constrained optimization problem:

$$\text{minimize} \quad \frac{1}{2}|\boldsymbol{w}|^2 \text{ subject to } y^{(i)}(\boldsymbol{w}^T \boldsymbol{x}^{(i)} + b) \geq 1, \quad i = 1, \ldots, m \qquad [4.18]$$

For nonlinearly separable data, SVM can be extended using *kernel functions*. Kernel functions allow SVM to operate in a higher dimensional feature space, where the data may become linearly separable. A common kernel function is the radial basis function (RBF) kernel used in many ML techniques:

$$K(\boldsymbol{x}, \boldsymbol{x}') = \exp(-\gamma|\boldsymbol{x} - \boldsymbol{x}'|^2), \quad [4.19]$$

where $\gamma > 0$ is a parameter that controls the shape of the decision boundary. The corresponding algorithm is given in Algorithm 4.4.

Algorithm 4.4. Support vector machine

Require: Training set $\{(\boldsymbol{x}^{(i)}, y^{(i)}), i = 1, \ldots, m\}$

Choose a kernel function $K(\cdot, \cdot)$ and its parameters

Formulate the constrained quadratic optimization problem:

$$\min_{\boldsymbol{w},b} \quad \frac{1}{2}\|\boldsymbol{w}\|^2$$

$$\text{s.t.} \quad y^{(i)}(\boldsymbol{w}^T\boldsymbol{x}^{(i)} + b) \geq 1, \quad i = 1, \ldots, m$$

Solve the optimization problem to obtain the optimal weight vector $\boldsymbol{w}$ and bias term b (or the dual variables in the kernel-based case)

for each new instance $\boldsymbol{x}$ **do**

Compute the decision function:

$f(\boldsymbol{x}) = \boldsymbol{w}^T\boldsymbol{x} + b$ (or the corresponding kernel-based version)

Classify the instance based on the sign of the decision function:

$$\hat{y} = \begin{cases} 1 & \text{if } f(\boldsymbol{x}) \geq 0 \\ -1 & \text{otherwise} \end{cases}$$

end for

It should be noted that SVMs can be extended to multi-class classification problems using the One-vs.-Rest (OvR) or One-vs.-One (OvO) approaches. In the OvR approach, a separate SVM is trained for each class against all the other classes. In the OvO approach, a separate SVM is trained for each pair of classes. In both cases, the class with the highest score or the most votes is chosen as the predicted class.

Figure 4.5 shows SVM classifications for the binary and multi-class classifications using the same datasets as for the k-NN and logistic regression models. The set up of the SVM is done as the following, listing also the available kernels that can be chosen:

```
from sklearn.svm import SVC

kernels = ['linear', 'poly','rbf','sigmoid']
k = 1
svmk = SVC(kernel=kernels[k])
```

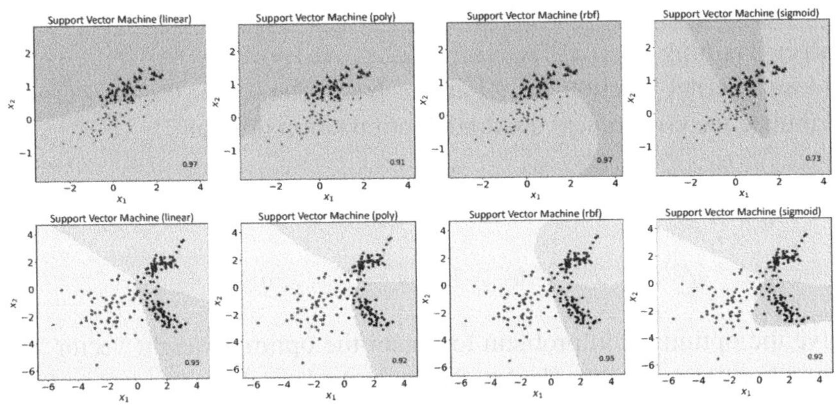

Figure 4.5. *Classification tasks using the support vector machines classifiers with various kernel functions. Upper panels: Binary classification with the same dataset and same coloring as in Figure 4.3. Lower panels: Multi-class classification with the same dataset as in Figure 4.4 and the same coloring. In both cases, boundary decisions are materialized by the background colors corresponding to the predictions*

The linear version performs similarly to the logistic regression with similar performances. The nonlinear kernels give rise to models with quite different performances and prediction as can be seen on the decision boundaries. It appears that the RBF kernel performs the best, while the sigmoid one seems not to be appropriate in the studied cases, either for binary or multi-class classifications. The decision in choosing a given kernel should be based on the dataset at hand.

4.2.4.4. *Improvements and ensemble techniques*

One significant improvement was brought by the so-called *bootstrapping*, which is a powerful re-sampling technique used in statistics and ML (Hastie et al. 2009). Given a dataset of size n, it consists of drawing a sample of size n with or without *replacement* (replacement means that the same data point can be picked up more than once) to infer descriptive statistics on this new sample. Repeated many times (each time drawing a new sample from the original dataset and computing the statistic), it results in a distribution of the computed statistic as an approximation of its sampling distribution. Bootstrapping is commonly used when it is difficult to make assumptions about the population or when the sample size is small. It allows us to estimate the uncertainty of our statistics without making any assumptions about the data distribution by creating multiple subsets of the original dataset, with replacement, and train a separate model on each subset. The final prediction is then made by aggregating the predictions of each model. This process helps to reduce variance and improve the model's robustness.

Ensemble techniques (Dong et al. 2020) enhance predictive performance compared to standard classification techniques as those considered above. More precisely, they increase generalizability and robustness of standard methods on which they are based (Hastie et al. 2009). The most common techniques are *bagging*, *boosting* and *stacking*, which are particularly effective at reducing overfitting by introducing diversity, reducing variance and therefore dealing with noise in the data. For instance, *random forest* (RF) is a popular technique based on bagging that uses decision trees as its base classifier (Biau and Scornet 2016). Decision trees recursively split the input space based on feature values to form a tree-like structure. Each node in the tree represents a decision rule based on a feature, and the leaf nodes represent the class labels. In addition to using bootstrap re-sampling for training, RF introduces an additional layer of randomness by selecting a random subset of features for splitting at each node. This process reduces the correlation between the trees, further improving the ensemble's performance.

4.3. AL in classification

AL is an approach where the learning algorithm actively queries the user (or an oracle) to provide labels for specific instances during the training process. This approach is particularly useful for classification tasks when labeled data are scarce or expensive to obtain, which is often the case in materials science, either experimentally of numerically, and geophysics. Then, AL aims to minimize the number of labeled examples required to achieve a desired level of performance, focusing on acquiring the most informative and useful examples for the learning process.

Various AL strategies have been proposed of which common ones include the following:

– *Uncertainty sampling*: in uncertainty sampling, the learner selects the instances for which it is least certain about the correct label. This is often quantified using the model's predicted class probabilities, with instances of the highest entropy or lowest maximum probability being chosen for labeling.

– *Query-by-committee*: in query-by-committee, multiple models (the committee) are trained on the current labeled dataset. The learner selects instances for which the committee members disagree the most, measured by the disagreement in their predicted labels or class probabilities.

– *Expected model change*: this strategy selects instances that are expected to have the largest impact on the current model if their true label was known. This can be quantified by estimating the expected change in the model parameters or loss function after incorporating the labeled instance.

– *Expected error reduction*: in this strategy, instances are selected based on their expected reduction in the model's generalization error. This can be estimated using techniques such as Monte Carlo sampling or Bayesian model averaging.

The typical AL workflow for classification tasks is described in Algorithm 4.5. After an initial training with given classification algorithms such as those described above, adding more instances in the training set is done until a criterion is reached. The latter is very problem dependent, but one has to bear in mind that the final dataset has to be as small as possible for a given performance or accuracy that is required. Therefore, a challenging task is to find a trade-off between exploring the space to discover new informative instances and exploiting the current knowledge to improve the model's performance.

Algorithm 4.5. Active learning workflow

Require: Initialize the labeled dataset with a small set of labeled instances, either selected randomly or using prior knowledge.
Train the classification model on the labeled dataset.
repeat
Select a batch of instances from the unlabeled dataset using the chosen active learning strategy.
Query the user to provide labels for the selected instances.
Update the labeled dataset with the newly labeled instances and retrain the model.
until a stopping criterion is met.

Other challenges come into play in AL. Model complexity is a bottleneck for AL, such as deep learning networks, due to their non-convex loss surfaces and high computational requirements for training and evaluation. AL assumes that the labeling is "perfect". In practice, labeling mechanisms may be prone to errors or inconsistencies, which can adversely affect the performance of the AL process.

A working example is illustrated in Figure 4.6. The bivariate Gaussian distribution dataset with $n = 600$ points (see Figure 4.3) was considered as a basis of the following simple implementation:

```
X_train, X_test, y_train, y_test = train_test_split(X, y,
    test_size= 0.2, random_state=42)
initial_idx = np.random.choice(range(len(X_train)),
    size=int(len(X_train) *0.1), replace=False)

X_pool = np.delete(X_train, initial_idx, axis=0)
y_pool = np.delete(y_train, initial_idx, axis=0)
X_initial = X_train[initial_idx]
y_initial = y_train[initial_idx]

model = SVC(kernel='rbf')
model.fit(X_initial, y_initial)
```

```
y_pred_test = model.predict(X_test)
initial_accuracy = accuracy_score(y_test, y_pred_test)

#Start the AL loop
n_queries = 40
for i in range(n_queries):
    query_idx = random.sample(range(len(X_pool)), 10)
    query_instances = X_pool[query_idx]
    y_pred = model.predict(query_instances)
    X_initial = np.append(X_initial, query_instances, axis=0)
    y_initial = np.append(y_initial, y_pred)
    X_pool = np.delete(X_pool, query_idx, axis=0)
    y_pool = np.delete(y_pool, query_idx, axis=0)
    model.fit(X_initial, y_initial)
    y_pred_test = model.predict(X_test)
    accuracy = accuracy_score(y_test, y_pred_test)
    y_pred_test = model.predict(X_test)
    accuracy = accuracy_score(y_test, y_pred_test)
```

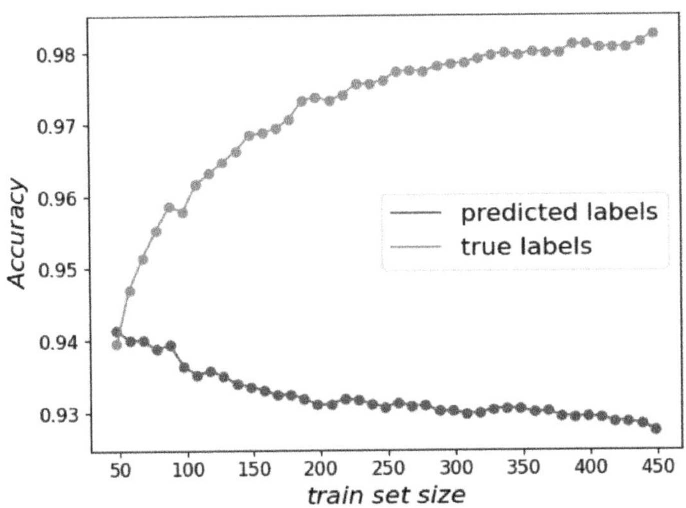

Figure 4.6. *Active learning for classification on a Gaussian bivariate distribution of $n = 600$ points with the same variance in two dimensions x_1 and x_2 used above*

The goal is to highlight through this basic AL implementation the importance as well as limitations that led to the various strategies listed above. The question that can be asked is the following: Are the 80% of the dataset needed to train the model? In other words, what is the minimum number of points required to reach the same accuracy? An AL simulation was performed with the SVM using the RBF kernel as a classifier. This SVM was shown to give the best performance for this dataset in the previous section.

An initial training set containing 10% of the full training set (80% of the dataset) is used to train a first SVM model. The process is iteratively refined by querying an additional 10 points at each iteration and retraining the Support Vector Machine (SVM) using their true labels, which are extracted from the training set. Typically, the queried data points are unlabeled, but, in this scenario, labels are sourced from the complete training set as though provided by an oracle. Figure 4.6 shows that the performance increases progressively in that case and reaches the one from the full training with a significantly lower number of data points (around 300 instead 480). The averaged accuracy curve shown was obtained by repeating 200 times this process.

Nevertheless, as mentioned above this is not the true story of AL. In a real scenario, a first model is trained with all the labeled data at hand, and increasing the performance of the model is done by adding new a priori unlabeled data for which a label has to be given by an oracle (like a human expert or a reference method to get "perfect" or "true" labels). Another idea could be to use the last trained model to predict the labels. This is precisely what is done for the "predicted labels" curve shown in Figure 4.6 that progressively worsen. This shows that the random sampling for new queries associated with prediction with the last trained model is not efficient. Both the wise sampling of new data points and the labeling are crucial and should be designed carefully, taking care for instance on the diversity in the input and output spaces (Wu et al. 2019; Jose et al. 2023).

4.3.1. *Application of classification*

An interesting application of classification in materials science is briefly mentioned here (Reinhart et al. 2017; Reinhart and

Panagiotopoulos 2018). This work aims at identifying and differentiating relevant ordered structures in data derived from molecular simulations or particle tracking. In contrast to other commonly used structure identification methods, this approach does not necessitate any predetermined description of the structures being sought. Feature vectors of structures are first build based on graph theory and adjacent matrices to generate impartial structural data that enable to measure the crystalline nature of particles located near defects, grain boundaries and interfaces. A multi-layer perceptron (see Chapters 2 and 3 in volume 2) classifies structural relationships among particles based on their local environmental topology. This method is used to categorize particles in a colloidal crystallization simulation and demonstrate the structural elements that go unnoticed when using conventional methods. Another work is based on the exploration of a refractory high entropy alloy Nb-Ti-Zr-Cr-Mo for mixture design. The composition space is explored to build an alloy library comprising hardness and ductility, two antagonistic properties, and to classify these compositions whether they are crystalline or amorphous (Garel et al. 2023).

4.4. Conclusion

Classification together with regression are the two pillars of supervised learning, with applications in many scientific fields, including but not limited to material science and geophysics. Classification techniques such as logistic regression, SVMs and ensemble methods were discussed in this chapter together with focusing on the most common ones for the sake of comprehensiveness.

Classification problems can benefit significantly from AL, a technique where the algorithm selectively queries the user (or some other information source) to obtain new data points considered the most informative. This is all the more true for regression problems, where AL is more emergent. In material science, AL can help in predicting properties of new, yet untested materials, thereby speeding up the discovery process at a minimal cost in time and financial means. In the same spirit in geophysics, it can aid in tasks like seismic monitoring or subsurface exploration, where obtaining labeled data can be costly or time consuming.

4.5. Acknowledgments

We acknowledge the CINES and IDRIS under Project No. INP2227/72914, as well as CIMENT/GRICAD for computational resources. This work was performed within the framework of the Centre of Excellence of Multifunctional Architectured Materials CEMAM-ANR-10-LABX-44-01 funded by the "Investments for the Future" Program. This work has been partially supported by MIAI@Grenoble Alpes (ANR-19-P3IA-0003). Discussions within the French collaborative network in high-temperature thermodynamics GDR CNRS3584 (TherMatHT) and in artificial intelligence in materials science GDR CNRS 2123 (IAMAT) are also acknowledged.

4.6. References

Biau, G. and Scornet, E. (2016). A random forest guided tour. *Test*, 25, 197–227.

Dong, X., Yu, Z., Cao, W., Shi, Y., Ma, Q. (2020). A survey on ensemble learning. *Frontiers of Computer Science*, 14, 241–258.

Garel, E., Parouty, J.-L., Van Landeghem, H., Verdier, M., Robaut, F., Coindeau, S., Boichot, R. (2023). Coupling mixture designs, high-throughput experiments and machine learning for accelerated exploration of multinary systems. *Materials & Design*, 231, 112055.

Hastie, T., Tibshirani, R., Friedman, J.H., Friedman, J.H. (2009). *The Elements of Statistical Learning: Data Mining, Inference, and Prediction*, vol. 2. Springer, New York.

Hey, T. and Trefethen, A. (2020). The fourth paradigm 10 years on. *Informatik Spektrum*, 42, 441–447.

Hey, T., Tansley, S., Tolle, K., Gray, J. (2009). *The Fourth Paradigm: Data-Intensive Scientific Discovery*. Microsoft Research, Redmond, WA [Online]. Available at: https://www.microsoft.com/en-us/research/publication/fourth-paradigm-data-intensive-scientific-discovery/.

Jose, A., de Mendonça, J.P.A., Devijver, E., Jakse, N., Monbet, V., Poloni, R. (2023). Regression tree-based active learning. *Data Mining and Knowledge Discovery*, 38, 1–41.

Karapiperis, K. and Andrade, J. (2021). Nonlocality in granular complex networks: Linking topology, kinematics and forces. *Extreme Mechanics Letters*, 42, 101041 [Online]. Available at: https://www.sciencedirect.com/science/article/pii/S2352431620302455.

Lookman, T., Balachandran, P.V., Xue, D., Yuan, R. (2019). Active learning in materials science with emphasis on adaptive sampling using uncertainties for targeted design. *npj Computational Materials*, 5(1), 21.

Rasmussen, C.E. and Williams, C.K. (2006). *Gaussian Processes for Machine Learning*, vol. 1. Springer, Cambridge.

Reinhart, W.F. and Panagiotopoulos, A.Z. (2018). Automated crystal characterization with a fast neighborhood graph analysis method. *Soft Matter*, 14(29), 6083–6089.

Reinhart, W.F., Long, A.W., Howard, M.P., Ferguson, A.L., Panagiotopoulos, A.Z. (2017). Machine learning for autonomous crystal structure identification. *Soft Matter*, 13(27), 4733–4745.

Schmidt, J., Marques, M.R., Botti, S., Marques, M.A. (2019). Recent advances and applications of machine learning in solid-state materials science. *npj Computational Materials*, 5(1), 83.

Shields, B.J., Stevens, J., Li, J., Parasram, M., Damani, F., Alvarado, J.I.M., Janey, J.M., Adams, R.P., Doyle, A.G. (2021). Bayesian reaction optimization as a tool for chemical synthesis. *Nature*, 590(7844), 89–96.

Wu, D., Lin, C.-T., Huang, J. (2019). Active learning for regression using greedy sampling. *Information Sciences*, 474, 90–105.

5

Artificial Neural Networks: Learning the Optimum Statistical Model from Data

Filippo GATTI

Laboratoire de Mécanique Paris-Saclay, UMR 9026, CentraleSupélec, CNRS, ENS Paris-Saclay, Université Paris-Saclay, Gif-sur-Yvette, France

The objective of this chapter is to provide a general background on machine learning basics to a non-expert readership with a strong background in computational science. Major theoretical aspects are outlined and supported by detailed examples and tutorials (along with the corresponding code snippets) to boost the learning process of both basic and advanced concepts with the help of practical and commented hands-on sessions, accessible to everybody. In particular, the chapter rephrases the machine learning problem according to an information theory paradigm that highlights the tight entanglement between two perspectives on data science: the statistic versus the optimization ones. Hereafter, sampling theory is firstly presented, including an

---

For a color version of all the figures in this chapter, see www.iste.co.uk/stefanou/machine1.zip.

Machine Learning in Geomechanics 1,
coordinated by Ioannis STEFANOU and Félix DARVE.

introduction on the Fisher approach to statistical learning and by an introduction to the principle of maximum entropy for conditional probability laws. The remainder of the chapter is devoted to present the optimization framework in which major machine learning problems are cast, consisting of minimizing the empirical risk over a class of parametric statistical models. This framework is presented in connection with the Fisher information and the MaxEnt principle presented at the beginning of the chapter. First- and second-order (stochastic) gradient descent optimization methods are presented in detail, with meaningful examples and hands-on sessions.

Among many sources, the chapter is largely inspired by Stéphane Mallat's Data Science lecture notes at Collège de France, from 2018 to 2022 (Campagne 2018, 2019, 2020, 2022)[1]. A special thanks goes to Jean-Eric Campagne, who redacted the lecture notes and took the time to review, comment and provide fruitful insights on this chapter. Another major source of inspiration of the chapter are several CentraleSupélec's engineering curriculum lecture notes, such as the *Statistics and Learning* lecture notes by Cournède (2020) and the lecture notes in *Modélisation des Incertitudes et Fiabilité des Ouvrages* by Clouteau (2022). The optimization section is inspired by Gabriel Peyré's *Course Notes on Optimization for Machine Learning* at École Normale Supérieure Paris (Paris Sciences et Lettres) (Peyré 2020)[2] and by Olga Mula's *Optimisation et programmation dynamique* lecture notes at Université Paris Dauphine (Mula 2019)[3].

5.1. Why `PyTorch`?

The AI revolution that we are living these days roughly started a decade ago, when AlexNet, a deep-learning model developed by (Krizhevsky et al. 2017), won the ImageNet Large Scale Visual

1 Those class notes are freely available on the Collège de France website: https://www.di.ens.fr/~mallat/CoursCollege.html

2 Available at: https://mathematical-tours.github.io/book-sources/optim-ml/OptimML.pdf.

3 Available at: https://www.ceremade.dauphine.fr/~mula/docs/teaching/optimM1/docs/notes_optim.pdf.

Recognition Challenge with a top-5 test error rate of 15.3%, compared to 26.2% achieved by the second-best entry. This milestone result proved that complex neural architectures can be now trained efficiently by crunching very large databases (in the case of AlexNet, 1.2 million high-resolution images in the ImageNet LSVRC-2010 contest, to be classified into 1,000 different classes) and achieving high accuracy. Five convolutional layers, some followed by max-pooling layers with a bottleneck of three fully connected layers and a final 1000-way softmax, for a total of $\approx$60 million parameters and 650,000 neurons, could now be efficiently trained to achieve highly accurate multi-class prediction, by leveraging an effective deployment on *graphic processing unit* (GPU). It was the beginning of the *deep-learning era*. By 2012, GPU programming had recently spread beyond the computer graphics realm, serving as generalized computing devices capable of racing against *central processing units* (CPUs) in general-purpose computing on GPU (GPUGP). Machine learning became one of the major GPUGP applications, followed by scientific image processing, linear algebra and statistics. With the introduction of the Nvidia GeForce 8 series, in 2006, and new generic stream processing units in the following years, it became clear that image processing algorithms (such as the one implemented in AlexNet) were particularly prone to efficient parallel GPU deployment, because of their incommensurably better performances when handling image-array operations: every piece of data, if shaped and interpreted as a suite of texture maps, can be efficiently filtered by swiping convolutional kernels across them. GPUGP (nowadays simply known as GPU) took advantage of powerful CPUs essentially because of their proneness to solve the same sequence of operations on a continuous stream of data, progressively offloaded from the host (the CPU, accessing large memory but at a slow rate) to the device (GPU, respectively). This offloading mechanism originated from the fact that GPUs were initially rather limited in terms of memory capacity (although this gap with CPU is progressively being remediated). To leverage the massively parallel computing capabilities, a GPU program must be tailored to efficiently handle the host-to-device offloading phase. However, machine learning algorithms, often cast into optimization problems, came at hand because of the introduction of stochastic gradient descent (SGD), which transformed the task of training deep neural architecture into a rather naive parallel problem (such as processing large databases of independent data realizations).

Therefore, in the immediate aftermath (roughly two years) of AlexNet's appearance, several libraries had sprouted already. Their goal is to enable common users with the power of deep learning, affordably trained on GPU devices, without being hindered by harsh CPU-GPU backend programming. Moreover, since 2016, *tensor processing units* (TPU) made their debut. TPUs are application-specific integrated circuits (ASIC) designed by Google with the aim to acclerate AI algorithms. Despite their inherent success, those deep-learning libraries have encountered alternate fortunes. Two major libraries survived the initial AI revolution and have established themselves as the most popular libraries for neural networks design and complex machine learning algorithms: `PyTorch` and `TensorFlow`. The development of those two libraries has been featured by several extra contributions of prototype libraries (`DistBelief`, `Caffe2`, `Theano`, `Keras` to cite a few) that have been progressively integrated into either one of two leading libraries, developed by Meta AI (developing `PyTorch`) and Google Brain Team (`TensorFlow`), respectively.

`PyTorch` was first released in 2016, originally developed by Adam Paszke, Sam Gross, Soumith Chintala and Gregory Chanan, as a `python` interface to the `Torch` open-source library, initially created by Idiap Research Institute and école polytechnique fédérale de Lausanne (EPFL) and migrated to `PyTorch` in 2017. `TensorFlow` was released in late 2015 by the Google Brain Team, based on its proprietary ancestor called `DistBelief`, started in 2011. The name `TensorFlow` was inspired by the archetype on which the library is based, that is, the multidimensional data array referred to as *tensors*. `TensorFlow` is specifically tailored for running on TPU running models on low floating-point precision (8-bit) rather than training them. Both `PyTorch` and `TensorFlow` are in their second versions, released in 2023 and 2019, respectively.

From a mere technical standpoint, the main difference between the two libraries is the way they assemble computational graphs. While `TensorFlow` distinguishes itself for using static graphs (compiled before running) to efficiently perform back-propagation, `PyTorch` adopts dynamic graphs that are assembled at run time. However, in its brand new version 2, recently released, `PyTorch` introduced an experimental tool to compile graphs. The difference between static and dynamic graphs is that static graphs are based on fixed tensor shapes and graph connections, which allows optimization of the graph operations before running it. However, certain optimization features cannot be applied to dynamic graphs assembled and modified at runtime, possibly reducing their overall speedup. Nevertheless, the `PyTorch`'s dynamic approach eases the coding part for non-expert users, since memory requirements (often unknown) need not be defined beforehand. Static graphs, on the contrary, allow secure model building with high-level API that enables portability on the cloud, in the browser, on-device, multiple CPUs and multiple GPUs or TPUs. Static graphs can be used by other supported languages, such as `C++` and `Java`, facilitating cross-platform deployment. On the contrary, `PyTorch` has its API for saving trained models in a serialized way and its API for distributed training on parallel architectures (`DistributedDataParallel` or `ModelParallel`).

`TensorFlow` dominated the AI scene at the very beginning (2012–2017) and rapidly left the step to `PyTorch` after release, due to its friendly interface and ease of code, which attracted new AI practitioners with poor background in machine learning and `python` coding in general. Therefore, the Google team released `TensorFlow` v2.0 in 2019, completely turning v1.0 upside down. The most intriguing novelty was the introduction of the "eager" mode, which introduced the possibility of switching from default static graph computation to the "Define-by-Run". Moreover, `TensorFlow` 2.0 performances on GPU drastically improved, enabling its use on that popular (compared to TPU at least) hardware. This allowed `TensorFlow` to regain a consistent portion of the market.

Very recently, in order to tackle larger machine learning problems, both `TensorFlow` and `PyTorch` have been accelerated by using the high-performance computing `Python` library named `JAX`[4], which is tailored for composable function transformations such as neural networks. `JAX` accelerates `Numpy` array computation and native `Python` functions by adopting Accelerated Linear Algebra (XLA) and by just-in-time compiling `NumPy` programs on GPUs and TPUs. `JAX` allows to program across multiple GPUs or TPU cores at once. Both grad and jit are instances of such transformations.

All in all, both `TensorFlow` and `PyTorch` are incredibly user-friendly and powerful machine learning libraries. `TensorFlow` is more mature, with advanced features for optimizing complex machine learning algorithms. However, `PyTorch` seemed more suitable for providing a first look at the fundamentals of neural networks.

5.2. Introduction to sampling theory

Neural networks (NN) represent a particular class of statistical models conceived to reproduce some complex hidden inference from a random input state $\boldsymbol{X} \in \mathcal{X}$ (the explanatory variable) to a random output state $\boldsymbol{Y} \in \mathcal{Y}$ (the explained variable). Without loss of generality, from now on it is assumed that:

– $\mathcal{X} \subset \mathbb{R}^{d_X}$ equipped with a Borel σ-algebra $\mathcal{B}\left(\mathbb{R}^{d_X}\right)$;

– $\mathcal{Y} \subset \mathbb{R}^{d_Y}$ with Borel σ-algebra $\mathcal{B}\left(\mathbb{R}^{d_Y}\right)$ in a *regression* framework (see section 2.1);

– $\mathcal{Y} \subset \mathbb{N}^m$ with the power set $\mathcal{P}\left(\mathbb{N}^m\right)$ as σ-algebra, in a *classification* framework (see section 3.3).

In the general *supervised learning*, $\mathcal{Y}$ and $\mathcal{X}$ differ, but a labeling function $f : \boldsymbol{x} \mapsto \boldsymbol{y} = f(\boldsymbol{x})$ can exist. This is the standard interpolation framework. If $\mathcal{Y} = \mathcal{X}$, the learning framework is *self-supervised*. *Unsupervised learning* deals with the sole explanatory

4 Available at: https://github.com/google/jax.

variable, which data aim at explaining, via clustering, discovering the factors of variations (Murphy 2022). To infer $\boldsymbol{Y}$ from $\boldsymbol{X}$, the focus is on the family of *conditional probability* laws. For a discrete random variable (i.e., absolutely continuous with respect to the count measure $K(\boldsymbol{dx})$) the conditional probability is provided by the Bayes formula:

$$\begin{aligned}&\left(P_{\boldsymbol{X}}.P_{\boldsymbol{Y}|\boldsymbol{X}=\cdot}\right)(A\times B)\\&=P_{(\boldsymbol{X},\boldsymbol{Y})}(A\times B)=\int_A P_{\boldsymbol{Y}|\boldsymbol{X}=\cdot}(\boldsymbol{x},B)\,p_{\boldsymbol{X}}(\boldsymbol{x})\,K(\boldsymbol{dx})=\\&=\sum_{\boldsymbol{x}_k\in D_0\cap A}P_{\boldsymbol{Y}|\boldsymbol{X}=\boldsymbol{x}}(B).P_{\boldsymbol{X}}(\{\boldsymbol{x}_k\})\quad\forall A\times B\in\mathcal{E}_{\mathcal{X}}\times\mathcal{E}_{\mathcal{Y}}\end{aligned}\qquad [5.1]$$

with $D_0=\{\boldsymbol{x}\in\mathcal{X}:\mathbb{P}(\boldsymbol{X}=\boldsymbol{x})>0\}$. However, equation [5.1] does not apply to other probability distributions due to the risk that the probability $P_{(\boldsymbol{X}\in A)}$ might equal zero (Billingsley 1995; Cournède 2020).

REMARK 5.1.– Discrete random variables $\boldsymbol{X}\in(\mathcal{X},\mathcal{P}(\mathcal{X}))$ are associated with the *count measure* K, that is, the point mass function $p(\boldsymbol{x})=\mathbb{P}(\{\boldsymbol{X}=\boldsymbol{x}\})$ such that $\forall A\in\mathcal{P}(\mathcal{X})$, $P_X(A)=\int_A p(\boldsymbol{x})\,K(\boldsymbol{dx})=\sum_{\boldsymbol{a}_k\in A}\mathbb{P}(\{\boldsymbol{X}=\boldsymbol{a}_k\})$. With the help of the so-called *Dirac's measure* $\sum_{\boldsymbol{a}_k\in A}\delta_{\boldsymbol{a}_k\in E}$ – with E being a countable set containing at least all $\boldsymbol{a}_k$ and such that $P_X(\{\boldsymbol{X}=\boldsymbol{a}_k\})>0$ – the Borel algebra can be adopted instead of $\mathcal{P}(\mathcal{X})$:

$$\begin{aligned}\forall A\in\mathcal{B}\left(\mathbb{R}^d\right)\quad P_X(A)&=\int_A p(\boldsymbol{x})\sum_{\boldsymbol{a}_k\in A}\delta_{\boldsymbol{a}_k\in E}(\boldsymbol{dx})\\&=\sum_{\boldsymbol{a}_k\in A}\mathbb{P}(\{\boldsymbol{X}=\boldsymbol{a}_k\})\end{aligned}$$

Therefore, the concept of conditional probability can be extended to a family of probability distributions defined as *transition kernels* $P_{\boldsymbol{Y}|\boldsymbol{X}=\cdot}:\mathcal{X}\times\mathcal{E}_{\mathcal{Y}}\to[0,1]$, such that $\forall\boldsymbol{x}\in\mathcal{X}\ P_{\boldsymbol{Y}|\boldsymbol{X}=\boldsymbol{x}}$ is a measure of probability on $\mathcal{E}_{\mathcal{Y}}$ and $\forall B\in\mathcal{E}_{\mathcal{Y}}\ P_{\boldsymbol{Y}\in B|\boldsymbol{X}=\cdot}$ is $\mathcal{E}_{\mathcal{X}}$-measurable. In this sense, it can be proven that $\forall A \times B \in \mathcal{E}_{\mathcal{X}} \times \mathcal{E}_{\mathcal{Y}}\ P_{(\boldsymbol{X},\boldsymbol{Y})}\left(A \times B\right)$ is an extension of the application $\left(P_{\boldsymbol{X}}.P_{\boldsymbol{Y}|\boldsymbol{X}=\cdot}\right)\left(A \times B\right)$ (Billingsley 1995; Cournède 2020):

$$\begin{aligned}\left(P_{\boldsymbol{X}}.P_{\boldsymbol{Y}|\boldsymbol{X}=\cdot}\right)\left(A \times B\right) &= P_{(\boldsymbol{X},\boldsymbol{Y})}\left(A \times B\right) \\ &= \int_A P_{\boldsymbol{Y}|\boldsymbol{X}=\cdot}\left(\boldsymbol{x}, B\right) P_{\boldsymbol{X}}\left(\boldsymbol{dx}\right) \qquad [5.2]\end{aligned}$$

If $P_{\boldsymbol{Y}|\boldsymbol{X}=\cdot}$ exists, the conditional transfer theorem (Billingsley 1995) states that for all measurable functions φ defined over $\mathcal{X} \times \mathcal{Y}$ and such that $P_{\varphi(\boldsymbol{X},\boldsymbol{Y})|\boldsymbol{X}=\cdot}$ exists, then $\forall \boldsymbol{x} \in \mathcal{X}$ $P_{\varphi(\boldsymbol{X},\boldsymbol{Y})|\boldsymbol{X}=\boldsymbol{x}} = P_{\varphi(\boldsymbol{x},\boldsymbol{Y})|\boldsymbol{X}=\boldsymbol{x}}$ and – as a corollary – $\mathbb{E}\left[\varphi\left(\boldsymbol{X},\boldsymbol{Y}\right)|\boldsymbol{X}=\boldsymbol{x}\right] = \mathbb{E}\left[\varphi\left(\boldsymbol{x},\boldsymbol{Y}\right)|\boldsymbol{X}=\boldsymbol{x}\right]$ (Cournède 2020). Finally, if the density $p\left(\boldsymbol{X},\boldsymbol{Y}\right)$ exists, then $\forall \boldsymbol{x} \in \mathcal{X}$ such that $p_{\boldsymbol{X}}\left(\boldsymbol{x}\right) > 0$, $p_{\boldsymbol{Y}|\boldsymbol{X}=\boldsymbol{x}}\left(\boldsymbol{y}\right) = \frac{p_{(\boldsymbol{X},\boldsymbol{Y})}(\boldsymbol{x},\boldsymbol{y})}{p_{\boldsymbol{X}}(\boldsymbol{x})}$ exists.

REMARK 5.2.– Following Billingsley (1995) and Cournède (2020), Fubini's theorem that is applied on any measurable function $\phi : \mathcal{X} \times \mathcal{Y} \rightarrow \mathbb{R}^{+}$ requires the notion of transition kernel:

$$\begin{aligned}&\int_{\mathcal{X}\times\mathcal{Y}} \phi\left(\boldsymbol{x},\boldsymbol{y}\right) P_{(\boldsymbol{X},\boldsymbol{Y})}\left(\boldsymbol{dx},\boldsymbol{dy}\right) \\ &= \int_{\mathcal{X}}\left(\int_{\mathcal{Y}} \phi\left(\boldsymbol{x},\boldsymbol{y}\right) P_{\boldsymbol{Y}|\boldsymbol{X}=\boldsymbol{x}}\left(\boldsymbol{dy}\right)\right) P_{\boldsymbol{X}}\left(\boldsymbol{dx}\right)\end{aligned}$$

If ϕ is integrable on $\left(\mathcal{X} \times \mathcal{Y}, \mathcal{E}_{\mathcal{X}} \times \mathcal{E}_{\mathcal{Y}}, P_{(\boldsymbol{X},\boldsymbol{Y})}\right)$, then the application $P_{\boldsymbol{X}}$-almost everywhere, $\boldsymbol{y} \mapsto \phi\left(\boldsymbol{x},\boldsymbol{y}\right)$ is $P_{\boldsymbol{Y}|\boldsymbol{X}=\boldsymbol{x}}$-integrable and $\boldsymbol{x} \mapsto \int_{\mathcal{Y}} \phi\left(\boldsymbol{x},\boldsymbol{y}\right) P_{\boldsymbol{Y}|\boldsymbol{X}=\boldsymbol{x}}\left(\boldsymbol{dy}\right)$ is $P_{\boldsymbol{X}}$-integrable (Cournède 2020). In particular, $\int_{\mathcal{Y}} P_{\boldsymbol{Y}|\boldsymbol{X}=\boldsymbol{x}}\left(\boldsymbol{dy}\right) = 1$, whereas $\int_{\mathcal{X}} P_{\boldsymbol{Y}|\boldsymbol{X}=\boldsymbol{x}}\left(B\right) P_{\boldsymbol{X}}\left(\boldsymbol{dx}\right) = P_{\boldsymbol{Y}}\left(B\right)$ is the *marginal* probability distribution.

A *NN* "learns" the hidden mapping between explanatory and explained states by unveiling some "characteristics"[5] of the unknown

5 A *characteristic* is defined as a mapping from $\mathcal{P}$, the set of probability measures on the measurable space $\left(\mathcal{X} \times \mathcal{Y}, \mathcal{E}_{\mathcal{X}} \times \mathcal{E}_{\mathcal{Y}}\right)$ into some measurable space $\left(\mathcal{Z}, \mathcal{E}_{\mathcal{Z}}\right)$.

data distribution $P_{(\boldsymbol{X},\boldsymbol{Y})}$, without any a priori knowledge on it, but only based on a limited amount of N random observations of the input and output states, collected into the dataset $\mathcal{D}_{\mathcal{XY}} = \{(\boldsymbol{x}_k, \boldsymbol{y}_k)\}_{k=1}^{N}$. This is the framework of a renewable experiment, with independent identically distributed samples $\mathcal{D}_{\mathcal{XY}}$, realizations associated with the unknown product law $P_{(\boldsymbol{X},\boldsymbol{Y})}^{\otimes N}$. The core idea is to employ those samples to construct *empirical statistics* on $\mathcal{D}_{\mathcal{XY}}$, defined as $S : \left((\mathcal{X} \times \mathcal{Y})^N, (\mathcal{E}_\mathcal{X} \times \mathcal{E}_\mathcal{Y})^{\otimes N}\right) \to (\mathcal{Z}, \mathcal{E}_\mathcal{Z})$, with $(\mathcal{Z}, \mathcal{E}_\mathcal{Z})$ being a measurable space and $\mathcal{Z}$ that can be $\mathcal{X} \times \mathcal{Y}$ or another functional space (Cournède 2020). Those empirical statistics help the *NN* to replicate the hidden characteristics of the unknown probability law, namely $\mathcal{G}(P_{(\boldsymbol{X},\boldsymbol{Y})})$, without any a priori presumption on it.

REMARK 5.3.– For $\mathcal{X} \in \mathbb{R}^{dx}$, a widely used statistics is the empirical average defined over a set of i.i.d. samples $\mathcal{D}_\mathcal{X} \in \left(\mathcal{X}^N, (\mathcal{E}_\mathcal{X})^{\otimes N}\right)$ and for any function $\boldsymbol{f} : (\mathcal{X}, \mathcal{E}_\mathcal{X}) \to \left(\mathbb{R}^{dx}, \mathcal{B}\left(\mathbb{R}^{dx}\right)\right)$:

$$S : \mathcal{D}_\mathcal{X} \mapsto \hat{\mathbb{E}}_{\boldsymbol{x} \in \mathcal{D}_\mathcal{X}} \left[\boldsymbol{f}\left(\boldsymbol{X}\right)\right] = \frac{1}{N}\sum_{k=1}^{N} \boldsymbol{f}\left(\boldsymbol{x}_k\right) \qquad [5.3]$$

The corresponding random variable is called $\bar{\boldsymbol{f}}_N = \frac{1}{N}\sum_{k=1}^{N} \boldsymbol{f}(\boldsymbol{X}_k)$ and it corresponds to the empirical estimation of the first-order moment $\boldsymbol{m}_1\left[\boldsymbol{f}\boldsymbol{X}\right]$. Note that:

$$\begin{aligned}\hat{\mathbb{E}}_{\boldsymbol{x} \in \mathcal{D}_\mathcal{X}} \left[\boldsymbol{f}\left(\boldsymbol{X}\right)\right] &= \int_X \boldsymbol{f}\left(\boldsymbol{x}\right) \hat{P}_{\boldsymbol{X}}\left(\boldsymbol{dx}\right) \\ &= \frac{1}{N}\sum_{k=1}^{N} \int_\mathcal{X} \boldsymbol{f}\left(\boldsymbol{x}\right) \delta_{\boldsymbol{x}_k}\left(\boldsymbol{dx}\right) = \frac{1}{N}\sum_{k=1}^{N} \boldsymbol{f}\left(\boldsymbol{x}_k\right)\end{aligned} \qquad [5.4]$$

The same holds for product empirical measures $\hat{P}\left(\boldsymbol{X},\boldsymbol{Y}\right) = \frac{1}{N}\sum_{k=1}^{N} \delta_{\boldsymbol{x}_k}\delta_{\boldsymbol{y}_k}$:

$$\hat{\mathbb{E}}_{(\boldsymbol{x},\boldsymbol{y})\in\mathcal{D}_{\mathcal{XY}}}\left[\boldsymbol{f}\left(\boldsymbol{X},\boldsymbol{Y}\right)\right]=\frac{1}{N}\sum_{k=1}^{N}\int_{\mathcal{X}\times\mathcal{Y}}\boldsymbol{f}\left(\boldsymbol{x},\boldsymbol{y}\right)\delta_{\boldsymbol{x}_k}(d\boldsymbol{x})\delta_{\boldsymbol{y}_k}(d\boldsymbol{y})$$
$$=\frac{1}{N}\sum_{k=1}^{N}\boldsymbol{f}(\boldsymbol{x}_k,\boldsymbol{y}_k) \qquad [5.5]$$

For conditional probability distributions, the empirical mean over $\boldsymbol{X}$ reads:

$$\hat{\mathbb{E}}_{\boldsymbol{x}\in\mathcal{D}_{\mathcal{X}}}\left[\boldsymbol{f}\left(\boldsymbol{Y}|\boldsymbol{X}\right)\right]=\frac{1}{N}\sum_{k=1}^{N}\boldsymbol{f}(\boldsymbol{Y}|\boldsymbol{x}_k) \qquad [5.6]$$

with $\hat{\mathbb{E}}_{\boldsymbol{x}\in\mathcal{D}_{\mathcal{X}}}\left[\boldsymbol{f}\left(\boldsymbol{Y}|\boldsymbol{X}\right)\right]$ being a random function of $\boldsymbol{Y}$. Taking the empirical mean over the joint product empirical measure leads to $\hat{\mathbb{E}}_{(\boldsymbol{x},\boldsymbol{y})\in\mathcal{D}_{\mathcal{XY}}}\left[\boldsymbol{f}\left(\boldsymbol{Y}|\boldsymbol{X}\right)\right]=\frac{1}{N}\sum_{k=1}^{N}\boldsymbol{f}\left(\boldsymbol{y}_k|\boldsymbol{x}_k\right)$.

Higher order empirical moments $\hat{\boldsymbol{m}}^{(p)}\left(\mathcal{D}_{\mathcal{X}}\right)$ are defined by adopting $\boldsymbol{f}$ as the multilinear application $\boldsymbol{f}:\boldsymbol{x}\mapsto\otimes_p\boldsymbol{x}$ with $\otimes_p\boldsymbol{x}$ p-order tensor in $\mathbb{R}^{pd_X}$. For instance, the empirical variance-covariance matrix (moment of order $p=2$) is defined by the following empirical statistics $S:\mathcal{D}_{\mathcal{X}}\mapsto\hat{\boldsymbol{m}}^{(2)}\left(\mathcal{D}_{\mathcal{X}}\right)-\hat{\boldsymbol{m}}^{(1)}\left(\mathcal{D}_{\mathcal{X}}\right)\otimes\hat{\boldsymbol{m}}^{(1)}\left(\mathcal{D}_{\mathcal{X}}\right)$:

$$S\left(\mathcal{D}_{\mathcal{X}}\right)=\hat{\mathbb{E}}_{\boldsymbol{x}\in\mathcal{D}_{\mathcal{X}}}\left[\boldsymbol{X}\otimes\boldsymbol{X}\right]-\bar{\boldsymbol{X}}_N\otimes\bar{\boldsymbol{X}}_N$$
$$=\hat{\mathbb{E}}_{\boldsymbol{x}\in\mathcal{D}_{\mathcal{X}}}\left[\left(\boldsymbol{X}-\bar{\boldsymbol{X}}_N\right)\otimes\left(\boldsymbol{X}-\bar{\boldsymbol{X}}_N\right)\right]$$

Another widely use statistic is the maximum defined as $S:\mathcal{D}_{\mathcal{X}}\mapsto\max_{x_i\in\mathcal{D}_{\mathcal{X}}}f(x_i)$.

The key concept to correctly employ empirical statistics is the *empirical measure* defined as $\hat{P}_{\boldsymbol{X}}=\frac{1}{N}\sum_{k=1}^{N}\delta_{\boldsymbol{x}_k}$. For instance, the notion of empirical measure allows us to define the empirical distribution function $\hat{F}$ that approximates the probability $P_X\left(]-\infty, x[\right)$ as the functional:

$$\hat{F}(\mathcal{D}_\mathcal{X}) : \mathcal{D}_\mathcal{X} \mapsto \left(\frac{1}{N}\sum_{k=1}^{N}\delta_{x_k}\right)\left(]-\infty, x[\right) = \frac{1}{N}\sum_{k=1}^{N}1_{]-\infty,x[}(x_k) \qquad [5.7]$$

Empirical measure allows to estimate the expectation operator by its empirical counterpart (see equation [5.3]):

$$\mathbb{E}_{\boldsymbol{x}\sim\hat{P}}\left(\boldsymbol{f}\left(\boldsymbol{X}\right)\right) = \int_{\mathcal{X}} \boldsymbol{f}\left(\boldsymbol{x}\right)\hat{P}\left(\boldsymbol{dx}\right) = \frac{1}{N}\sum_{k=1}^{N}\boldsymbol{f}\left(\boldsymbol{x}_k\right) \qquad [5.8]$$

The same goes for other statistics. Therefore, any sought characteristics $\mathcal{G}(P_{(\boldsymbol{X},\boldsymbol{Y})})$ can be approached by its empirical counterpart $\mathcal{G}(\hat{P}_{(\boldsymbol{X},\boldsymbol{Y})})$, provided that some sort of convergence of the estimator for N sufficiently large is granted (by the law of large numbers).

The empirical product measure $\hat{P}_{(\boldsymbol{X},\boldsymbol{Y})} \quad : \quad \left((\mathcal{X}\times\mathcal{Y})^N,\right.$ $\left.(\mathcal{E}_\mathcal{X}\times\mathcal{E}_\mathcal{Y})^{\otimes N}\right) \to (\mathcal{X}\times\mathcal{Y}, \mathcal{E}_\mathcal{X}\times\mathcal{E}_\mathcal{Y})$ reads:

$$\hat{P}_{(\boldsymbol{X},\boldsymbol{Y})} = \frac{1}{N}\sum_{k=1}^{N}\delta_{\boldsymbol{x}_k}\cdot\delta_{\boldsymbol{y}_k} \qquad [5.9]$$

One tricky aspect of this empirical approach is that for continuous random variables $\boldsymbol{X} \in \mathcal{X} \subset \mathbb{R}^{dx}$, $\hat{P}_{(\boldsymbol{X},\boldsymbol{Y})}$ is not continuous with respect to the Lebesgue's measure μ over $\mathbb{R}^{dx}$: $\hat{P}_{(\boldsymbol{X},\boldsymbol{Y})}(\{\boldsymbol{x}_k, \boldsymbol{y}_k\}) = \frac{1}{N}$ but $\mu\left\{\boldsymbol{x}_k, \boldsymbol{y}_k\right\} = 0$. In this case, determining an empirical density $\hat{p}_{\boldsymbol{X}}$ is not possible, since the Radon–Nikodym theorem does not apply. However, $\hat{p}_{\boldsymbol{Y}|\boldsymbol{X})}$ is usually approximated by the Monte Carlo approach, that is, by fitting the histograms of the $\boldsymbol{y}_k$ obtained for each $\boldsymbol{x}_k$ (Murphy 2022), or by the kernel method (Cournède 2020). The conditional probability $\hat{P}_{\boldsymbol{Y}|\boldsymbol{X}=\cdot}$, which is of paramount importance to make the NN learn the general stochastic mapping from $\boldsymbol{X}$ to $\boldsymbol{Y}$, is instead modeled with a statistical model. This approach is called a Monte Carlo approximation to the distribution.

5.2.1. *Statistical models and maximum likelihood estimator*

NN infer $\boldsymbol{y}$ from $\boldsymbol{x}$ through a trainable mapping $\boldsymbol{h}_{\boldsymbol{\theta}} : \mathcal{X} \to \mathcal{Y}$, associated with a parametric probability distribution $P_{\boldsymbol{\theta}}(\boldsymbol{Y}|\boldsymbol{X} = \cdot) = P(\boldsymbol{Y}|\boldsymbol{X} = \cdot, \boldsymbol{\theta})$, which defines the so-called *statistical model.* Statistical models are families of (conditional) probability distributions depending on a set of parameters $\boldsymbol{\theta}$. In this case, a statistical model[6]:

$$\mathcal{H}_{\Theta} := \left\{P_{\boldsymbol{\theta}}(\boldsymbol{Y}|\boldsymbol{X}), \boldsymbol{Y} = \boldsymbol{h}_{\boldsymbol{\theta}}(\boldsymbol{X}), \boldsymbol{\theta} \in \Theta \subset \mathbb{R}^{d_{\Theta}}\right\} \qquad [5.10]$$

$P_{\boldsymbol{\theta}}(\boldsymbol{Y}|\boldsymbol{X})$ is tailored to approach the true probability law, once it has been postulated that $P \in \mathcal{H}_{\Theta}$. In this case, there would exist $\boldsymbol{\theta}^* \in \Theta$ such that $P_{\boldsymbol{\theta}^*} = P$. Parametric statistical models are often chosen to be dominated by a σ-finite measure μ and sufficiently regular (see remark 5.5). This allows us to define a probability density $p_{\boldsymbol{\theta}}(\boldsymbol{Y}|\boldsymbol{X})$. The training process is performed based on the available realizations $\mathcal{D}_{\mathcal{X}\mathcal{Y}}$, thus adopting the empirical probability $\hat{P}$. In other words, the key idea in frequentist statistics is to view the data $\mathcal{D}_{\mathcal{X}\mathcal{Y}}$ as a random variable (different datasets of size N can be drawn), and the parameters from which the data are drawn, $\boldsymbol{\theta}^*$, as a fixed but unknown constant (Murphy 2022).

REMARK 5.4.– A very well-known regular statistical model is the set of MultiVariate Normals (MVNs) defined as the set of d-variate normal distributions (Nielsen 2023):

$$\mathcal{H}_{\Theta} = \left\{p_{\boldsymbol{\theta}}(\boldsymbol{X}) = \mathcal{N}(\boldsymbol{\mu}, \boldsymbol{\Sigma}), \boldsymbol{\theta} = \{\boldsymbol{\mu}, \boldsymbol{\Sigma}\} \in \mathbb{R}^d \times \mathbb{S}_d^{+*}(\mathbb{R})\right\} \qquad [5.11]$$

6 Θ is considered a finite-dimensional space, although non-parametric statistical models associated with infinite-dimensional Θ exist.

with $\mathcal{S}_d^{+*}(\mathbb{R})$ being the set of square symmetric real-valued positive-definite matrices of dimension $d \times d$. The dimension of the statistical model $\dim(\mathcal{H}_\Theta) = d_\Theta$ is equal to $d_\Theta = \frac{d(d+3)}{2}$ and the associated probability density function reads:

$$p_{\boldsymbol{\theta}}(\boldsymbol{x}) = \left((2\pi)^d \cdot \det(\boldsymbol{\Sigma})\right)^{-\frac{1}{2}} e^{-\frac{1}{2}\|\boldsymbol{\Sigma}^{-\frac{1}{2}}(\boldsymbol{x}-\boldsymbol{\mu})\|^2} \qquad [5.12]$$

$\boldsymbol{\mu}$ and $\boldsymbol{\Sigma}$ can depend on $\boldsymbol{X}$ via linear or nonlinear mappings. Usually, the covariance $\boldsymbol{\Sigma}$ is fixed and does not depend on the input data, according to the *homoscedastic* regression hypothesis. If $\boldsymbol{\Sigma} = \boldsymbol{\Sigma}(\boldsymbol{X})$, the model is called *heteroscedastic* (Murphy 2022).

MVN distributions are tensorized normal distributions of univariate normal distributions $p_{\boldsymbol{\theta}}(x_k) = (2\pi\sigma_k^2)^{-\frac{1}{2}} e^{-\frac{x_k-\mu_k}{2\sigma_k^2}}$, with $\sigma_k^2 > 0$ being the kth eigenvalue of $\boldsymbol{\Sigma}$.

The MVN are a particular case of the *exponential* family, characterized by following probability density parameterized by $\boldsymbol{\theta}$:

$$p_{\boldsymbol{\theta}}(\boldsymbol{x}) = h(\boldsymbol{x}) \exp\left(\sum_{k=1}^{d_\Theta} S_k(\boldsymbol{x}) \cdot \phi_k(\boldsymbol{\theta}) - \ln Z(\boldsymbol{\theta})\right) \qquad [5.13]$$

with the S_k being sufficient statistics, $h(\boldsymbol{x})$ a scale constant and $Z(\boldsymbol{\theta}) = \int_{\mathcal{X}} h(\boldsymbol{x}) \exp\left(\sum_{k=1}^{d_\Theta} S_k(\boldsymbol{x})\, \phi_k(\boldsymbol{\theta})\right) \mu(\boldsymbol{dx})$ a normalization constant called the *partition function*. Parametric probability distributions $p_{\boldsymbol{\theta}}(\boldsymbol{Y}|\boldsymbol{X})$ following the exponential family can be obtained by replacing ϕ_k in equation [5.13] by a mapping $(\boldsymbol{x}, \boldsymbol{\theta}) \mapsto \phi_k(\boldsymbol{x}, \boldsymbol{\theta})$.

REMARK 5.5.– A *regular* model $\mathcal{H}_\Theta = \left\{P_{\boldsymbol{\theta}}(\boldsymbol{X}), \boldsymbol{\theta} \in \Theta \subset \mathbb{R}^{d_\Theta}\right\}$ is a parametric dominated (by the σ-finite measure μ, either Lebesgue or count measure) statistical model, described therefore by the parametric family of probability density functions $p_{\boldsymbol{\theta}}(\boldsymbol{x})$ and that verifies the following regularity conditions (Cournède 2020):

C1 Θ is an open set and $\forall \boldsymbol{x} \in \mathcal{X}$ and $\forall (\boldsymbol{\theta}, \boldsymbol{\theta}^{'}) \in \Theta^2$,

$$p_{\boldsymbol{\theta}}(\boldsymbol{x}) > 0 \iff p_{\boldsymbol{\theta}'}(\boldsymbol{x}) > 0$$

This implies that all $p_{\boldsymbol{\theta}} \in \mathcal{H}_\Theta$ have the same support denoted $\text{supp}(p_{\boldsymbol{\theta}})$, $\forall \theta \in \Theta$.

C2 $\forall \boldsymbol{\theta} \in \Theta$, $p_{\boldsymbol{\theta}}$ can be differentiated under the integral[7]:

$$\boldsymbol{\nabla}_\theta \int_{\text{supp}(p_{\boldsymbol{\theta}})} p_{\boldsymbol{\theta}} \cdot \mu(\boldsymbol{dx}) = \int_{\text{supp}(p_{\boldsymbol{\theta}})} \boldsymbol{\nabla}_\theta p_{\boldsymbol{\theta}} \cdot \mu(\boldsymbol{dx}) \qquad [5.15]$$

C3 $p_{\boldsymbol{\theta}} \in \mathcal{C}^2(\Theta)$;

C4 $\forall \boldsymbol{\theta} \in \Theta$, $\forall i, j,\ 1 \leq i, j \leq p$ the theorem of derivation under integral can be applied as follows[8]:

$$\boldsymbol{\nabla}_{\boldsymbol{\theta}} \otimes \int_{\text{supp}(p_{\boldsymbol{\theta}})} \boldsymbol{\nabla}_\theta p_{\boldsymbol{\theta}}(\boldsymbol{X})\, \mu(\boldsymbol{dx})$$

$$= \int_{\text{supp}(p_{\boldsymbol{\theta}})} \boldsymbol{\nabla}_{\boldsymbol{\theta}} \otimes \boldsymbol{\nabla}_\theta p_{\boldsymbol{\theta}}(\boldsymbol{X})\, \mu(\boldsymbol{dx}) \qquad [5.16]$$

MVNs are regular.

If $\mathcal{H}_\Theta$ is chosen to be *identifiable*, that is, if the map $\boldsymbol{\theta} \mapsto P_{\boldsymbol{\theta}}$ is injective, then $\boldsymbol{\theta}^*$ is unique. Moreover, if all $P_{\boldsymbol{\theta}} \in \mathcal{H}_\Theta$ are dominated by a σ-finite measure (i.e. they are absolutely continuous with respect to a Lebesgue measure or count measure μ), they admit densities $p_{\boldsymbol{\theta}}$[9]. In this framework, since it has been postulated that $\exists \boldsymbol{\theta}^* \in \Theta$ such that $P_{(\boldsymbol{X},\boldsymbol{Y})} = P_{\boldsymbol{\theta}^*}$ (or its conditional counterpart, equivalently), any of the

7 According to (Billingsley 1995), C2 holds if the gradient is locally dominated by an integrable function g, that is, there exists a neighborhood $\text{supp}(p_{\boldsymbol{\theta}})$ of θ and g such that $\int_{\text{supp}(p_{\boldsymbol{\theta}})} g(x)\mu(dx) < +\infty$ such that almost everywhere on a neighborhood V of $\boldsymbol{x}$:

$$\left|\frac{\partial p_{\boldsymbol{\theta}}}{\partial \theta_k}\right| \leq g. \qquad [5.14]$$

8 Under the same conditions that led to equation [5.14]

9 In a Bayesian framework, $\boldsymbol{\theta}$ can be considered as a random vector too: in this case, $p_{\boldsymbol{\theta}}$ must be interpreted as a conditional probability $p(\boldsymbol{Y}|\boldsymbol{X}, \boldsymbol{\theta})$ (Cournède 2020).

k characteristics can be considered as a function $g_k(\boldsymbol{\theta}) = \mathcal{G}_k(P_{\boldsymbol{\theta}})$. A scheme of the overall learning process of a statistical model is shown in Figure 5.1.

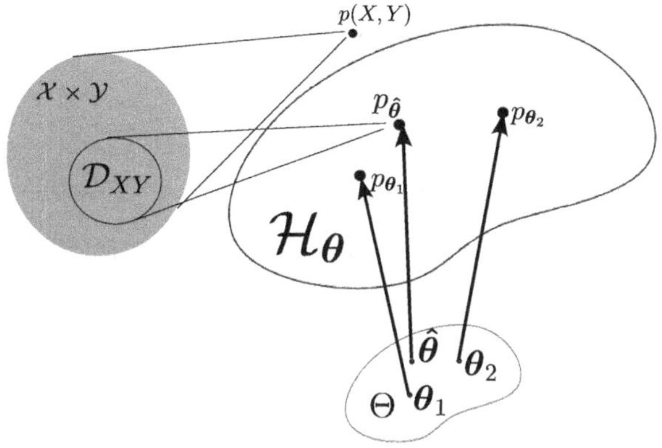

Figure 5.1. *Schematic view of the learning framework of a statistical model*

A first approach is provided by the so-called *substitution method*, where the set of characteristics $\boldsymbol{g}^* = \mathcal{G}(P)$ is approached by the characteristics $\boldsymbol{g}(\boldsymbol{\theta}) = \mathcal{G}(P_{\boldsymbol{\theta}})$, computed adopting the empirical measure ($\hat{\boldsymbol{g}}(\boldsymbol{\theta})$), by minimizing the distance $\min_{\boldsymbol{\theta}\in\Theta} d(\hat{\boldsymbol{g}}(\boldsymbol{\theta}), \boldsymbol{g}^*)$ (for the choice of the metric; see the inspiring work by Nielsen (2023), for further insights). The substitution method is a generalization of the *method of moments* that employs the statistics provided by the first d_Θ conditional moments of the distribution $\hat{\boldsymbol{g}}(\boldsymbol{\theta}) = (S_1(\boldsymbol{\theta}; \mathcal{D}_{\mathcal{XY}}), \ldots, S_{d_\Theta}(\boldsymbol{\theta}; \mathcal{D}_{\mathcal{XY}}))$, such that $\hat{g}_k(\boldsymbol{\theta}) = g_k^*$ with:

$$\hat{g}_k(\boldsymbol{\theta}) = S_k(\boldsymbol{\theta}; \mathcal{D}_{\mathcal{XY}}) = \hat{\boldsymbol{m}}^{(k)}_{\mathbf{Y}|\mathbf{X}}(\mathcal{D}_{\mathcal{XY}}) = \hat{\mathbb{E}}_{(\boldsymbol{x},\boldsymbol{y})\in\mathcal{D}_{\mathcal{XY}}}\left[\bigotimes_k \boldsymbol{h}_{\boldsymbol{\theta}}(\boldsymbol{X})\right] \quad [5.17]$$

$$g_k^* = \mathbb{E}_{\boldsymbol{x}\sim p}\left[\mathbb{E}_{\boldsymbol{y}\sim p(\cdot|\boldsymbol{x})}\left(\bigotimes_k \boldsymbol{Y}\right)\right]$$

to obtain a system of d_Θ equations in d_Θ unknowns (Cournède 2020). The latter statement corresponds to the choice of g as a *sufficient* statistics (Nielsen 2023).

As an alternative, Fisher (1922) proposed to estimate $\boldsymbol{\theta}^*$ as the maximum-likelihood estimator (MLE), a statistic defined as:

$$\hat{\boldsymbol{\theta}}_N\left(\mathcal{D}_{\mathcal{X}\mathcal{Y}}\right)=\underset{\boldsymbol{\theta}\in\Theta}{\arg\max}\prod_{k=1}^{N}p_{\boldsymbol{\theta}}\left(\boldsymbol{y}_k|\boldsymbol{x}_k\right) \qquad [5.18]$$

with $L_{\boldsymbol{Y}|\boldsymbol{X}}\left(\boldsymbol{\theta};\mathcal{D}_{\mathcal{X}\mathcal{Y}}\right)=\prod_{k=1}^{N}p_{\boldsymbol{\theta}}\left(\boldsymbol{y}_k|\boldsymbol{x}_k\right)$ being a statistics called *empirical conditional likelihood.* The latter corresponds to the conditional probability of a data set of i.i.d. realizations $\mathcal{D}_{\mathcal{X}\mathcal{Y}}$ so that their joint probability density reads[10]:

$$p_{\boldsymbol{\theta}}\left(\boldsymbol{X},\boldsymbol{Y};\mathcal{D}_{\mathcal{X}\mathcal{Y}}\right)=\prod_{k=1}^{N}p_{\boldsymbol{\theta}}\left(\boldsymbol{x}_k,\boldsymbol{y}_k\right)=\prod_{k=1}^{N}p_{\boldsymbol{\theta}}\left(\boldsymbol{y}_k|\boldsymbol{x}_k\right)\cdot p_{\boldsymbol{\theta}}\left(\boldsymbol{x}_k\right) \qquad [5.19]$$

and the conditional probability density reads:

$$p_{\boldsymbol{\theta}}\left(\boldsymbol{Y}|\boldsymbol{X};\mathcal{D}_{\mathcal{X}\mathcal{Y}}\right)=\frac{\prod_{k=1}^{N}p_{\boldsymbol{\theta}}\left(\boldsymbol{x}_k,\boldsymbol{y}_k\right)}{\int_{\mathcal{Y}}p_{\boldsymbol{\theta}}\left(\boldsymbol{x},\boldsymbol{y}\right)\mu(\boldsymbol{dy})}=\prod_{k=1}^{N}p_{\boldsymbol{\theta}}\left(\boldsymbol{y}_k|\boldsymbol{x}_k\right) \qquad [5.20]$$

$\hat{\boldsymbol{\theta}}_N(\mathcal{D}_{\mathcal{X}\mathcal{Y}})$ must be intended as a random variable depending on the random sample $\mathcal{D}_{\mathcal{X}\mathcal{Y}}$ of size N and its distribution is known as the *sampling distribution* of the estimator. A distribution of estimates is obtained by *pushing* the data distribution through the empirical conditional likelihood (Murphy 2022). The choice of the likelihood estimation is motivated by the fact that $\hat{\boldsymbol{\theta}}_N$ is a (weakly) *consistent* estimator for i.i.d. samples, as the following theorem states (Cournède 2020; Campagne 2022):

10 In a Bayesian framework, $p_{\boldsymbol{\theta}}\left(\boldsymbol{X},\boldsymbol{Y}\right)=p_{\boldsymbol{\theta}}\left(\boldsymbol{X},\boldsymbol{Y}|\boldsymbol{\theta}\right)=p_{\boldsymbol{\theta}}\left(\boldsymbol{Y}|\boldsymbol{X};\boldsymbol{\theta}\right)\cdot p_{\boldsymbol{\theta}}\left(\boldsymbol{X}|\boldsymbol{\theta}\right)$. The *full log-likelihood* corresponds to $\ell_{(\boldsymbol{X},\boldsymbol{Y})}\left(\boldsymbol{\theta}\right)=\ln L_{(\boldsymbol{X},\boldsymbol{Y})}\left(\boldsymbol{\theta}\right)=\ln p_{\boldsymbol{\theta}}\left(\boldsymbol{X},\boldsymbol{Y}\right)=\ln p_{\boldsymbol{\theta}}\left(\boldsymbol{X},\boldsymbol{Y}|\boldsymbol{\theta}\right)$, whereas the *marginal log-likelihood* corresponds to $\ln\mathcal{L}_{\boldsymbol{X}}\left(\boldsymbol{\theta}\right)=\ln p_{\boldsymbol{\theta}}\left(\boldsymbol{X}|\boldsymbol{\theta}\right)$.

THEOREM 5.1.– Given a sequence of i.i.d. random variables $\mathcal{D}_{\mathcal{XY}} = \{(\boldsymbol{x}_k, \boldsymbol{y}_k)\}_{k=1}^{N}$, then:

$$\forall(\boldsymbol{\theta}, \boldsymbol{\theta}^*) \in \Theta, \boldsymbol{\theta} \neq \boldsymbol{\theta}^*, \forall\varepsilon > 0$$

$$\lim_{N\to+\infty} \mathbb{P}\left[p_{\boldsymbol{\theta}}\left(\mathcal{D}_{\mathcal{XY}}\right) - p_{\boldsymbol{\theta}^*}\left(\mathcal{D}_{\mathcal{XY}}\right) < \varepsilon\right] = 1$$

Moreover, the optimal estimator $\hat{\boldsymbol{\theta}}_N$ does not change if the problem is recast into the minimization of the so-called *conditional negative log-likelihood* $NLL_{\mathbf{Y}|\mathbf{X}}\left(\boldsymbol{\theta}; \mathcal{D}_{\mathcal{XY}}\right)$, that is, by taking the negative natural logarithm of $p_{\boldsymbol{\theta}}\left(\boldsymbol{Y}|\boldsymbol{X}\right)$ and minimizing the following function instead:

$$NLL_{\mathbf{Y}|\mathbf{X}}\left(\boldsymbol{\theta}; \mathcal{D}_{\mathcal{XY}}\right) = -\sum_{k=1}^{N} \ln p_{\boldsymbol{\theta}}\left(\boldsymbol{y}_k|\boldsymbol{x}_k\right) \qquad [5.21]$$

with $\ell\left(\boldsymbol{x}_k, \boldsymbol{y}_k\right)\left(\boldsymbol{\theta}\right) = \ln p_{\boldsymbol{\theta}}\left(\boldsymbol{y}_k|\boldsymbol{x}_k\right)$. Figure 5.2 shows an example of the likelihood function for a parametric Bernoulli's distribution.

REMARK 5.6.– The average conditional negative log-likelihood $\frac{1}{N}NLL_{\mathbf{Y}|\mathbf{X}}$ associated with the univariate Laplace distribution, namely, $Y|X \sim$Laplace$\left(h_{\boldsymbol{\theta}_\mu}(X), \sigma\right)$ with $\boldsymbol{\theta} = \left\{\boldsymbol{\theta}_\mu, \sigma\right\}$ reads:

$$NLL_{Y|X}\left(\boldsymbol{\theta}; \mathcal{D}_{\mathcal{XY}}\right) = \ln\left(2\sigma\right) + \frac{1}{N\sigma}\sum_{k=1}^{N} |y_k - h_{\boldsymbol{\theta}_\mu}(x_k)| \quad [5.22]$$

It is well known that the variance of the Laplace distribution is equal to $\mathbb{V}_{y\sim\text{Laplace}\left(h_{\boldsymbol{\theta}_\mu}(X),\sigma\right)} = 2\theta_\sigma^2$, with σ called *diversity*. In equation [5.27], the variance is fixed and independent of the input for *homoscedastic* regression (Murphy 2022). Forcing $\boldsymbol{\nabla}_{\boldsymbol{\theta}} NLL_{Y|X} = \mathbf{0}$ allows to find the optimum diversity $\hat{\sigma}_N$:

$$\hat{\sigma}_N = \frac{1}{N}\sum_{k=1}^{N} |y_k - h_{\boldsymbol{\theta}_{\hat{\mu}_N}}\left(x_k\right)|$$

and $h_{\boldsymbol{\theta}_{\hat{\mu}_N}}(x_k)$ representing the per-sample average. If the statistical model was unconditioned and with only two parameters (sufficient statistics), then $h_{\boldsymbol{\theta}_{\mu}}(X) = \mu$ and therefore its optimum value is obtained by equating the derivative $\frac{\partial NLL_{\mathbf{Y}|\mathbf{X}}}{\partial \mu}(\hat{\mu}_N, \hat{\sigma}_N; \mathcal{D}_{\mathcal{X}}) = -\sum_{k=1}^{N} sgn\,(y_k - \hat{\mu}_N) = 0$. The latter derivative exists everywhere except for samples $y_k = \hat{\mu}_N$ and it is null if the y_k are ordered:

$$\hat{\mu}_N = \text{med}(Y; \mathcal{D}_{\mathcal{X}\mathcal{Y}})$$

with med being the sample median. If N is odd, $\text{med}(Y; \mathcal{D}_{\mathcal{X}\mathcal{Y}}) = y_{\frac{N+1}{2}}$. As a matter of fact, the function $\sum_{k=1}^{N} |y_k - \hat{\mu}_N|$ is continuous everywhere and it decreases for $\mu <\text{med}(Y; \mathcal{D}_{\mathcal{X}\mathcal{Y}})$ and increases for $\mu >\text{med}(Y; \mathcal{D}_{\mathcal{X}\mathcal{Y}})$. The values of $\hat{\mu}_N$ and $\hat{\sigma}_N$ are matching the empirical moments of the probability density, as assumed in equation [5.17], for the method of moments.

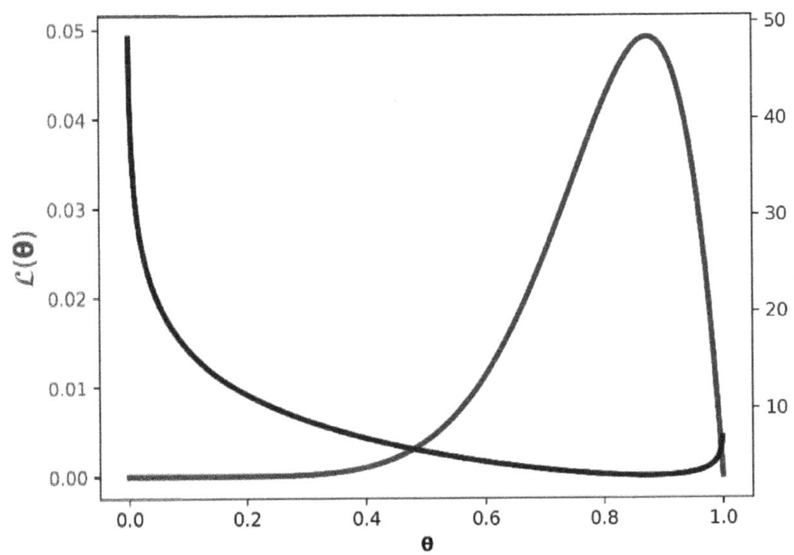

Figure 5.2. *Bernoulli distribution* $p_{\boldsymbol{\theta}}(x, |\mathcal{D}_{\mathcal{X}}) =$*Ber*$(x|\boldsymbol{\theta}, \mathcal{D}_{\mathcal{X}}) = \prod_{k=1}^{N} \theta^{1_{x_k=1}} (1-\theta)^{1_{x_k=0}}$

Another interesting example is the average conditional $\frac{1}{N}NLL_{\boldsymbol{Y}|\boldsymbol{X}}$ of a generic probability density within the MVN family in equation [5.13], that is, for $p_{\boldsymbol{\theta}}(\boldsymbol{Y}|\boldsymbol{X}) = \mathcal{N}\left(\boldsymbol{h}_{\boldsymbol{\mu}}(\boldsymbol{X}), \boldsymbol{\Sigma}\right)$. with $\boldsymbol{\theta} = \{\boldsymbol{\mu}, \boldsymbol{\Sigma}\}$ and N samples reads:

$$\frac{1}{N}NLL_{\mathbf{Y}|\mathbf{X}}(\boldsymbol{\theta}) = \frac{d}{2}\ln(2\pi) + \frac{1}{2}\ln\det\boldsymbol{\Sigma} + \frac{1}{2N}\sum_{k=1}^{N}\|\boldsymbol{\Sigma}^{-\frac{1}{2}}\left(\boldsymbol{y}_k - \boldsymbol{h}_{\boldsymbol{\mu}}(\boldsymbol{x}_k)\right)\|^2 \quad [5.23]$$

The optimum values of $\boldsymbol{\theta}$ are obtained for $\boldsymbol{\nabla}_{\boldsymbol{\theta}}NLL_{\boldsymbol{Y}|\boldsymbol{X}} = \mathbf{0}$ (see example 5.1 for the implementation of the algorithm to find the optimum parameter $\hat{\boldsymbol{\theta}}_N$). Under the homoscedasticity hypothesis, $\hat{\boldsymbol{\Sigma}}_N$ corresponds to the empirical covariance matrix:

$$\hat{\boldsymbol{\Sigma}}_N = \frac{1}{N}\sum_{k=1}^{N}\left[\left(\boldsymbol{y}_k - \boldsymbol{h}_{\hat{\boldsymbol{\mu}}_N}\right)\otimes\left(\boldsymbol{y}_k - \boldsymbol{h}_{\hat{\boldsymbol{\mu}}_N}(\boldsymbol{x}_k)\right)\right]$$

In other words, in analogy with the method of moments in equation [5.17], $\hat{\boldsymbol{\Sigma}}_N$ corresponds to the empirical covariance of the observations $\boldsymbol{y}$ provided an "mean" model $\boldsymbol{h}_{\hat{\boldsymbol{\mu}}_N}(\boldsymbol{X})$, obtained by minimizing the mean square error (MSE) $\frac{1}{N}\|\boldsymbol{y}_k - \boldsymbol{h}_{\boldsymbol{\mu}}(\boldsymbol{x}_k)\|^2$. As seen in equation [2.3], $\boldsymbol{h}_{\boldsymbol{\mu}}(\boldsymbol{X})$ is often a linear model (linear regression set up), which generically reads $\boldsymbol{h}_{\boldsymbol{\mu}}(\boldsymbol{x}) = \boldsymbol{W}\boldsymbol{x} + \boldsymbol{b}$. In this case, the overall set of parameters to adjust is composed of the weights $\boldsymbol{W}$, the "bias" $\boldsymbol{b}$ and $\boldsymbol{\Sigma}$. If the model was simply $\boldsymbol{h}_{\boldsymbol{\mu}}(\boldsymbol{x}) = \boldsymbol{W}\boldsymbol{x}$, section 2.2 in Chapter 2 showed that the optimum weight vector is equal to:

$$\hat{\boldsymbol{W}}_N = \left(\mathbf{X}_T\mathbf{X}\right)^{-1}\boldsymbol{X}_T\boldsymbol{Y}$$

with $\mathbf{X} = \sum_{k=1}^{N}\boldsymbol{x}_k\otimes\boldsymbol{e}_k$ and $\mathbf{Y} = \sum_{k=1}^{N}\boldsymbol{Y}_k\otimes\boldsymbol{e}_k$. If $\boldsymbol{h}_{\boldsymbol{\mu}}(\boldsymbol{x}) = \boldsymbol{b}$ minimizing the $NLL_{\boldsymbol{Y}|\boldsymbol{X}}$ leads to find the optimum bias:

$$\hat{\boldsymbol{b}}_N = \frac{1}{N}\sum_{k=1}^{N}\boldsymbol{y}_k$$

Moreover, the model can become linear and heteroskedastic if one constructs the dependency of $\boldsymbol{\Sigma}$ on $\boldsymbol{x}$.

EXAMPLE 5.1 (*NLL_X* **minimization with** `PyTorch`).– In the following example, the maximum likelihood criterion is implemented in `PyTorch` to find the parameters of the statistical model:

$$\mathcal{H}_\Theta := \left\{ p_{\boldsymbol{\theta}} = \frac{1}{(2\pi\theta_\sigma^2)^{\frac{1}{2}}} e^{-\frac{(x-\theta_\mu)^2}{2\cdot\theta_\sigma^2}} \right\}$$

and:

$$\mathcal{H}_\Theta := \left\{ p_{\boldsymbol{\theta}} = \frac{1}{2\theta_\sigma} e^{-\frac{|x-\theta_\mu|}{\theta_\sigma}} \right\}$$

both with parameters:

$$\Theta := \{\theta_\mu, \theta_\sigma\}, \quad \theta_\mu \in \mathbb{R}, \theta_\sigma \in \mathbb{R}^+$$

The two probability distributions in $\mathcal{H}_\Theta$ correspond to normal and Laplace distributions, respectively. The "true" probability distribution belongs to $\mathcal{H}_\Theta$ and it is set by selecting two parameters:

$$\bar{\theta}_\mu = \mu_X, \quad \bar{\theta}_\sigma = \sigma_X$$

Therefore, $p_{\bar{\theta}_\mu, \bar{\theta}_\sigma} = p_{\mu_X, \sigma_X} \in \mathcal{H}_\Theta$, which implies that $\hat{\boldsymbol{\theta}}_N \approx \boldsymbol{\theta}^*$. The random variable X generates the data set of i.i.d. realizations $\mathcal{D}_X = \{x_k\}_{k=1}^N$, with N =10,000. For the sake of simplicity, instead of maximizing the empirical average log-likelihood defined, the code minimizes the empirical $\mathcal{NLL}[\boldsymbol{X}]$ so to find:

$$\begin{bmatrix} \hat{\theta}_\mu \\ \hat{\theta}_\sigma \end{bmatrix} = \underset{\boldsymbol{\theta} \in \Theta}{\arg\min} \sum_{k=1}^N \ln \frac{1}{p_{\boldsymbol{\theta}}(\boldsymbol{x}_k)} \quad [5.24]$$

which still satisfies the convergence in probability expressed in theorem 5.1 that reads:

$$\forall \boldsymbol{\theta} \neq \boldsymbol{\theta}^*, \forall \varepsilon > 0$$

$$\lim_{N \to +\infty} \mathbb{P}\left[\frac{1}{N} \ln p_{\boldsymbol{\theta}}\left(\mathcal{D}_{\mathcal{X}}\right) < \varepsilon + \frac{1}{N} p_{\boldsymbol{\theta}^*}\left(\mathcal{D}_{\mathcal{X}}\right)\right] = 1 \qquad [5.25]$$

In particular, given the statistical model in this example, the $\mathcal{NLL}$ function for a univariate normal distribution reads:

$$NLL_X(\boldsymbol{\theta}; \mathcal{D}_{\mathcal{X}}) = \frac{N}{2} \ln(2\pi) + N \ln \theta_\sigma + \frac{1}{2\theta_\sigma^2} \sum_{i=1}^{N} (x_i - \theta_\mu)^2 \qquad [5.26]$$

and for a Laplace distribution:

$$NLL_X(\boldsymbol{\theta}; \mathcal{D}_{\mathcal{X}}) = N(\ln \theta_\sigma + \ln 2) + \frac{1}{\theta_\sigma} \sum_{i=1}^{N} |x_i - \theta_\mu| \qquad [5.27]$$

The minimization is achieved by using the `AdamW` optimizing algorithm (see section 5.3.7.5 and https://pytorch.org/docs/stable/generated/torch.optim.AdamW.html), over n_e = 100,000 epochs.

The code below minimizes the NLL_X to find the "best" estimators $\hat{\theta}_\mu \approx \bar{\theta}_\mu = \mu_X$ and $\hat{\theta}_\sigma \approx \bar{\theta}_\sigma = \sigma_X$. The result of the iterative minimization is depicted in Figure 5.3.

```
# Source: https://stackoverflow.com/questions/72469496/
#how-to-use-pytorch-for-maximum-likelihood-estimation-with
#-restrict-optimization
import numpy as np
import torch
from matplotlib import pyplot as plt
```

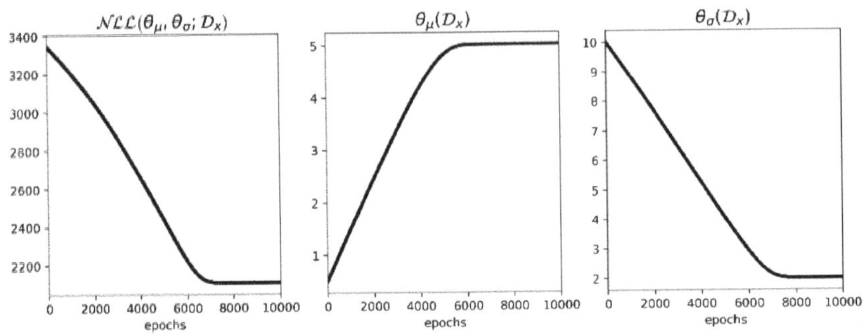

Figure 5.3. *Iterative minimization (with `AdamW` algorithm) of $\mathcal{NLL}(\theta_\mu, \theta_\sigma; \mathcal{D}_\mathcal{X})$. The convergence of the "best" parameters $(\hat{\theta}_\mu, \hat{\theta}_\sigma)$ toward the "true" parameters $(\bar{\theta}_\mu, \bar{\theta}_\sigma)$ is shown, with respect to the iteration number (epoch). The "true" probability distribution corresponds to $\mathcal{N}(\mu_X, \sigma_X)$, with μ_X = 5 and σ_X = 2. The figure was generated using the code*

```
'''
Learning the parameters of an MVN probability distribution
by minimize its empirical NLL
'''
# Fix the pseudo-random generator see to grant
reproducibility torch.manual_seed(0)
# Create a dataset of nX samples from normal distribution
N = 10000
# real mean
mu_X = torch.tensor(np.array([5.0]),
                    dtype=torch.float64,
                    requires_grad=False).tile((N,))
# real standard deviation
sigma_X = torch.tensor(np.array([2.0]),
                    dtype=torch.float64,
                    requires_grad=False)

# generate random samples from p=N(mu_X, sigma_X)
D_X = torch.normal(mean=mu_X, std=sigma_X)
# Initialize the values of the estimators
theta_mu = torch.tensor(np.array([0.5]),
                        dtype=torch.float64,
                        requires_grad=True)
D_X.requires_grad = False
```

```
theta_sigma = torch.tensor(np.array([10.0]),
                           dtype=torch.float64,
                           requires_grad=True)

# Define the optimizer
eta = 0.0001
optimizer = torch.optim.AdamW([theta_mu, theta_sigma],
            lr = eta)

# Number of training epochs
n_e = 100000

# Minimize the Negative Log-Likelihood iteratively
track_nll=[]
track_theta_mu = []
track_theta_sigma = []
for epoch in range(n_e):
  # Compute the negative log-likelihood
  nll = N*(0.5*np.log(2.0*np.pi)+theta_sigma.log())
  nll+= ((((D_X-theta_mu)/theta_sigma).pow(2))/2.0).sum()
  # Reset the optimizer parameters
  optimizer.zero_grad()
  # Compute the gradient
  nll.backward()

  if epoch % 100 == 0:
    print("NLL: {}; theta_mu: {}; theta_sigma: {}".
                             format(nll.data.numpy(),
                             theta_mu.data.numpy(),
                             theta_sigma.data.numpy()))

  # Apply delta rule
  optimizer.step()

  # Update tracking record
  track_nll.append(float(nll))
  track_theta_mu.append(float(theta_mu))
  track_theta_sigma.append(float(theta_sigma))

# plot convergence curves
fig, ax = plt.subplots(nrows=1, ncols=3, sharex=True,
          figsize=(12,4))
```

```
ax[0].plot(track_nll,
           color='k',
           linewidth=3,
           label=r"$\mathcal{NLL}$")
ax[1].plot(track_theta_mu,
           color='k',
           linewidth=3,
           label=r"$\theta_\mu$")
ax[2].plot(track_theta_sigma,
           color='k',
           linewidth=3,
           label=r"$\theta_\sigma$")
ax[0].set_xlim(0,n_e)
ax[0].set_xlabel("epochs")
ax[1].set_xlabel("epochs")
ax[2].set_xlabel("epochs")
ax[0].set_title(r"$\mathcal{NLL}\left(\theta_\mu,
        \theta_\sigma;\mathcal{D}_x\right)$", fontsize=15)
ax[1].set_title(r"$\theta_\mu\left(\mathcal{D}_x\right)$",
                fontsize=15)
ax[2].set_title(r"$\theta_\sigma\left(\mathcal{D}_x\right)$",
                fontsize=15)
fig.savefig("NLL_gauss.png", dpi=300, bbox_inches="tight")
```

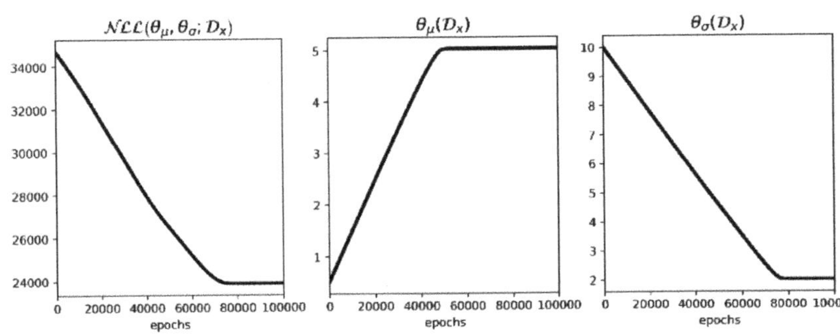

Figure 5.4. *Iterative minimization (with* `AdamW` *algorithm) of $\mathcal{NLL}(\theta_\mu, \theta_\sigma; \mathcal{D}_X)$. The convergence of the "best" parameters $(\hat{\theta}_\mu, \hat{\theta}_\sigma)$ toward the "true" parameters $(\bar{\theta}_\mu, \bar{\theta}_\sigma)$ is shown, with respect to the iteration number (epoch). The "true" probability distribution corresponds to Laplace(μ_X, σ_X), with $\mu_X = 5$ and $\sigma_X = 2$. The figure was generated using the code*

```
# Source: https://stackoverflow.com/questions/72469496/
#how-to-use-pytorch-for-maximum-likelihood-estimation-with
#-restrict-optimization
import numpy as np
import torch
from matplotlib import pyplot as plt

'''
Laplace probability distribution
'''
from torch.distributions.laplace import Laplace
# Fix the pseudo-random generator see to grant
# reproducibility
torch.manual_seed(0)
# Create a dataset of nX samples from normal distribution
N = 10000
# real mean
mu_X = torch.tensor(np.array([5.0]),
                    dtype=torch.float64,
                    requires_grad=False).tile((N,))
# real standard deviation
sigma_X = torch.tensor(np.array([2.0]),
                    dtype=torch.float64,
                    requires_grad=False)

# generate random samples from p=N(mu_X, sigma_X)
generator = Laplace(mu_X, sigma_X)
D_X = generator.sample()
D_X.requires_grad = False

# Initialize the values of the estimators
theta_mu = torch.tensor(np.array([0.5]),
                        dtype=torch.float64,
                        requires_grad=True)
theta_sigma = torch.tensor(np.array([10.0]),
                        dtype=torch.float64,
                        requires_grad=True)

# Define the optimizer
learning_rate = 0.0001
optimizer = torch.optim.AdamW([theta_mu, theta_sigma],
                        lr = learning_rate)

n_e = 100000
```

```
# Minimize the Negative Log-Likelihood iteratively
track_nll=[]
track_theta_mu = []
track_theta_sigma = []
for epoch in range(n_e):
  # Compute negative log-likelihood
  nll = N*(np.log(2.0)+theta_sigma.log())
  nll+= (D_X-theta_mu).abs().sum()/theta_sigma.abs()
  optimizer.zero_grad()
  nll.backward()


  if epoch % 100 == 0:
    print("NLL: {}; theta_mu: {}; theta_sigma: {}".
                              format(nll.data.numpy(),
                              theta_mu.data.numpy(),
                              theta_sigma.data.numpy()))
  optimizer.step()
  track_nll.append(float(nll))
  track_theta_mu.append(float(theta_mu))
  track_theta_sigma.append(float(theta_sigma))

# plot convergence curves
fig, ax = plt.subplots(nrows=1, ncols=3, sharex=True,
            figsize=(12,4))
ax[0].plot(track_nll,
           color='k',
           linewidth=3,
           label=r"$\mathcal{NLL}$")
ax[1].plot(track_theta_mu,
           color='k',
           linewidth=3,
           label=r"$\theta_\mu$")
ax[2].plot(track_theta_sigma,
           color='k',
           linewidth=3,
           label=r"$\theta_\sigma$")
ax[0].set_xlim(0,n_e)
ax[0].set_xlabel("epochs")
ax[1].set_xlabel("epochs")
ax[2].set_xlabel("epochs")
ax[0].set_title(r"$\mathcal{NLL}\left(\theta_\mu,
                \theta_\sigma;\mathcal{D}_x\right)$",
                 fontsize=15)
```

```
ax[1].set_title(r"$\theta_\mu\left(\mathcal{D}_x\right)$",
                fontsize=15)
ax[2].set_title(r"$\theta_\sigma\left(\mathcal{D}_x\right)$",
                fontsize=15)
fig.savefig("NLL_laplace.png", dpi=300, bbox_inches="tight")
```

The interest of adopting $NLL_{\boldsymbol{Y}|\boldsymbol{X}}\left(\boldsymbol{\theta};\mathcal{D}_{\mathcal{XY}}\right)$ resides in the fact that it corresponds to the empirical average of *Kullback–Leibler divergence* between true and modeled probability densities (see Figure 5.1). In other words, the MLE $\hat{\boldsymbol{\theta}}_N\left(\mathcal{D}_{\mathcal{XY}}\right)$ in equation [5.26] represents the optimum choice to achieve the *minimum discrimination information* (MDI), which consists of empirically minimizing the dissimilarity between $P\left(\boldsymbol{Y}|\boldsymbol{X}\right)$ and $P_{\boldsymbol{\theta}}\left(\boldsymbol{Y}|\boldsymbol{X}\right)$, quantified by the so-called (empirical) *Kullback–Leibler divergence* between the conditional probabilities. For absolutely continuous probability laws, the latter is defined as[11,12]:

$$\mathbb{D}_{KL}\left[P\left(\boldsymbol{Y}|\boldsymbol{X}\right):P_{\boldsymbol{\theta}}\left(\boldsymbol{Y}|\boldsymbol{X}\right)\right]=\mathbb{E}_{(\boldsymbol{x},\boldsymbol{y})\sim p}\left[\ln\frac{p\left(\boldsymbol{Y}|\boldsymbol{X}\right)}{p_{\boldsymbol{\theta}}\left(\boldsymbol{Y}|\boldsymbol{X}\right)}\right]\geq 0 \quad [5.28]$$

The Kullback–Leibler divergence is positive[13] and it vanishes when $P_{\boldsymbol{\theta}}$ approaches P-almost everywhere. Therefore, in analogy with the substitution method, $\mathbb{D}_{KL}\left[P\left(\boldsymbol{Y}|\boldsymbol{X}\right):P_{\boldsymbol{\theta}}\left(\boldsymbol{Y}|\boldsymbol{X}\right)\right]$ can be seen as an asymmetric "statistical distance" between the two characteristics

---

11 In Nielsen (2021), the original $\mathbb{D}_{KL}\left[p:q\right]$ definition for two probability distributions P and Q defined over the measurable space $\left(\mathcal{X},\mathcal{E}_{\mathcal{X}}\right)$ reads:

$$\mathbb{D}_{KL}\left[P:Q\right]=\int_{\mathcal{X}}\ln\left(\frac{dP}{dQ}\right)dP=\int_{\mathcal{X}}\frac{dP}{dQ}\cdot\ln\left(\frac{dP}{dQ}\right)dQ \quad P\ll Q$$

and $\mathbb{D}_{KL}\left[P:Q\right]=+\infty$ if $P\not\ll Q$. Moreover, if $P\ll Q$ and both P and Q satisfy the Radon–Nykodym theorem w.r.t. a σ finite measure μ such that their densities $p=\frac{dP}{d\mu}$ and $q=\frac{dq}{d\mu}$ exist, then

$$\mathbb{D}_{KL}\left[P:Q\right]=\int_{\mathcal{X}}p(\boldsymbol{x})\cdot\ln\left(\frac{p\left(\boldsymbol{x}\right)}{q\left(\boldsymbol{x}\right)}\right)\mu(\boldsymbol{dx})$$

12 $\mathbb{D}_{KL}\left[p:q\right]$ is defined if and only if supp$q\subset$ supp$q$. As a convention, $0\cdot\ln 0=0\cdot\ln\frac{0}{0}=0$.

13 Because of the Jensen inequality with $\ln x$ being a strictly concave function for $x>0$.

$g^* = -\ln p(\boldsymbol{Y}|\boldsymbol{X})$ and $\hat{g}(\boldsymbol{\theta}) = NLL_{\boldsymbol{Y}|\boldsymbol{X}}(\boldsymbol{\theta};\mathcal{D}_{\mathcal{XY}})$ and the minimization of the Kullback–Leibler divergence corresponds to minimizing the mean "distance" $\mathbb{E}_{(\boldsymbol{x},\boldsymbol{y})\sim P}[\hat{g}(\boldsymbol{\theta}) - g^*]$. In fact, equation [5.28] can be rewritten as:

$$\begin{aligned} &\mathbb{D}_{KL}\left[P(\boldsymbol{Y}|\boldsymbol{X}) : P_{\boldsymbol{\theta}}(\boldsymbol{Y}|\boldsymbol{X})\right] \\ &= -H_d(\boldsymbol{Y}|\boldsymbol{X}) - \mathbb{E}_{(\boldsymbol{x},\boldsymbol{y})\sim P}\left[\ln p_{\boldsymbol{\theta}}(\boldsymbol{Y}|\boldsymbol{X})\right] \end{aligned} \qquad [5.29]$$

with $H_d(P(\boldsymbol{Y}|\boldsymbol{X})) = -\mathbb{E}_{(\boldsymbol{x},\boldsymbol{y})\sim P}[\ln p(\boldsymbol{Y}|\boldsymbol{X})]$ being Shannon's conditional differential entropy, that is, the average amount of information needed to describe the outcome of a continuous random variable $\boldsymbol{Y}$ given that the value of another continuous random variable $\boldsymbol{X}$ is known. $H_d(P(\boldsymbol{Y}|\boldsymbol{X}))$ does not depend on $\boldsymbol{\theta}$ whatsoever and it can be considered, from now on, as a constant regardless of underlying true yet unknown probability distribution from which the data are sampled. Therefore, to achieve the MDI and minimize $\mathbb{D}_{KL}\left[P(\boldsymbol{Y}|\boldsymbol{X}) : P_{\boldsymbol{\theta}}(\boldsymbol{Y}|\boldsymbol{X})\right]$, we need to minimize the second term in the right-hand side of equation [5.29], whose empirical version equals the average $NLL_{\boldsymbol{Y}|\boldsymbol{X}}$:

$$\begin{aligned} &\hat{\mathbb{E}}_{\boldsymbol{x}\in\mathcal{D}_{\mathcal{X}}}\left[\mathbb{D}_{KL}\left[P(\boldsymbol{Y}|\boldsymbol{x}) : P_{\boldsymbol{\theta}}(\boldsymbol{Y}|\boldsymbol{x})\right]\right] \\ &= -\hat{\mathbb{E}}_{(\boldsymbol{x},\boldsymbol{y})\in\mathcal{D}_{\mathcal{XY}}}\left[\ln p_{\boldsymbol{\theta}}(\boldsymbol{Y}|\boldsymbol{X})\right] \\ &-\frac{1}{N}\sum_{k=1}^{N}\ln p_{\boldsymbol{\theta}}(\boldsymbol{y}_k|\boldsymbol{x}_k) = \frac{1}{N}NLL_{\boldsymbol{Y}|\boldsymbol{X}}(\boldsymbol{\theta};\mathcal{D}_{\mathcal{XY}}) \end{aligned} \qquad [5.30]$$

REMARK 5.7.– Note that $\mathbb{D}_{KL}\left[P : P_{\boldsymbol{\theta}}\right] \neq \mathbb{D}_{KL}\left[P_{\boldsymbol{\theta}} : P\right]$.

EXAMPLE 5.2.– (**Compute** $\mathbb{D}_{KL}\left[P : P_{\boldsymbol{\theta}}\right]$ **with** `PyTorch`)[14]

To avoid underflow issues when computing this quantity, this loss expects the argument input in the log space. As with the other losses in `PyTorch`, this function expects the first argument, input, to be the

14 Available at: https://pytorch.org/docs/stable/generated/torch.nn.KLDivLoss.html.

output of the model (e.g. the neural network) and the second, target, to be the observations in the dataset.

```
from torch import nn
import torch.nn.functional as F
# loss = loss_pointwise.sum() / input.size(0)
kl_loss = nn.KLDivLoss(reduction="batchmean")
# input should be a distribution in the log space
ptheta = F.log_softmax(torch.randn(3, 5,
          requires_grad=True), dim=1)
# Sample a batch of distributions. Usually this would come
# from the dataset
p = F.softmax(torch.rand(3, 5), dim=1)
output = kl_loss(ptheta, p)

# alternatively
kl_loss = nn.KLDivLoss(reduction="batchmean",
            log_target=True)
log_target = F.log_softmax(torch.rand(3, 5), dim=1)
output = kl_loss(ptheta, log_target)
```

5.2.2. *On the MLE optimization problem*

From an optimization perspective, $\hat{\boldsymbol{\theta}}_N\left(\mathcal{D}_{\mathcal{XY}}\right)$ defined in equation [5.18] does not necessarily exist, nor it is unique. If $\hat{\boldsymbol{\theta}}_N \in \mathring{\Theta}$ and C1 condition holds, $p_{\boldsymbol{\theta}}$ is differentiable with respect to $\boldsymbol{\theta}$ and a necessary condition for an estimator $\hat{\boldsymbol{\theta}}_N$ to exist is the following[15]:

$$\nabla_{\boldsymbol{\theta}} NLL_{\mathbf{Y}|\mathbf{X}}\left(\hat{\boldsymbol{\theta}}_N;\mathcal{D}_{\mathcal{XY}}\right) = -\hat{\boldsymbol{s}}_{\mathbf{Y}|\mathbf{X}}\left(\hat{\boldsymbol{\theta}}_N;\mathcal{D}_{\mathcal{XY}}\right) = \mathbf{0} \quad [5.31]$$

with $\hat{\boldsymbol{s}}_{\mathbf{Y}|\mathbf{X}}\left(\hat{\boldsymbol{\theta}}_N;\mathcal{D}_{\mathcal{XY}}\right) = \sum_{k=1}^{N} \nabla_{\boldsymbol{\theta}} \ln p_{\boldsymbol{\theta}}\left(\boldsymbol{y}_k|\boldsymbol{x}_k\right) = \mathbf{0}$ the empirical conditional *score*, generically defined as $\boldsymbol{s}_{\mathbf{Y}|\mathbf{X}}\left(\boldsymbol{\theta}\right) =$

15 The gradient of the conditional likelihood expressed in equation [5.20], for i.i.d. samples, reads:

$$\hat{\boldsymbol{s}}_{\mathbf{Y}|\mathbf{X}}\left(\boldsymbol{\theta};\mathcal{D}_{\mathcal{XY}}\right) = \sum_{k=1}^{N} \frac{\nabla_{\boldsymbol{\theta}} p_{\boldsymbol{\theta}}\left(\boldsymbol{y}_k|\boldsymbol{x}_k\right)}{p_{\boldsymbol{\theta}}\left(\boldsymbol{y}_k|\boldsymbol{x}_k\right)} = \frac{\nabla_{\boldsymbol{\theta}} p_{\boldsymbol{\theta}}\left(\boldsymbol{Y}|\boldsymbol{X};\mathcal{D}_{\mathcal{XY}}\right)}{p_{\boldsymbol{\theta}}\left(\boldsymbol{Y}|\boldsymbol{X};\mathcal{D}_{\mathcal{XY}}\right)}$$

$\boldsymbol{\nabla}_{\boldsymbol{\theta}} \ln p_{\boldsymbol{\theta}} (\boldsymbol{Y}|\boldsymbol{X})$. The empirical conditional score drives the optimization algorithm until the optimal solution is reached. In other words, while repeatedly sampling data from the unknown true distribution and computing the conditional score, its mean value tends to zero asymptotically ($N \rightarrow +\infty$). If conditions C1 and C2 hold for the statistical model (differentiability under integral[16]), the true conditional score $\boldsymbol{s}_{\boldsymbol{Y}|\boldsymbol{X}} (\boldsymbol{\theta}^*)$ satisfies the Bartlett first identity[17]:

$$\mathbb{E}_{\boldsymbol{x} \sim p_{\boldsymbol{\theta}^*}} \left[\mathbb{E}_{\boldsymbol{y} \sim p_{\boldsymbol{\theta}^*}(\cdot|\boldsymbol{x})} \left[\boldsymbol{s}_{\boldsymbol{Y}|\boldsymbol{X}=\boldsymbol{x}} (\boldsymbol{\theta}^*) \right] \right] = \mathbf{0} \qquad [5.32]$$

In other words, from a frequentist point of view, the true conditional score is a centered random vector or unbiased estimator. In an empirical framework, equation [5.32] corresponds to the optimum condition $\hat{\boldsymbol{s}} (\boldsymbol{\theta}^*; \mathcal{D}_{\mathcal{XY}}) = \mathbf{0}$[18], obtained by adopting i.i.d. samples. Moreover, if $NLL_{\boldsymbol{Y}|\boldsymbol{X}} (\boldsymbol{\theta}; \mathcal{D}_{\mathcal{XY}})$ is convex and Θ a convex set, the

16 The Leibniz integral rule is granted by the C2 condition. Thanks to the property of the gradient of the function $\boldsymbol{\theta} \mapsto \ln f(\boldsymbol{\theta})$, with $f : \Theta \rightarrow \mathbb{R}_*^+$ and such that $\boldsymbol{\nabla}_{\boldsymbol{\theta}} \ln f(\boldsymbol{\theta}) = \frac{\boldsymbol{\nabla}_{\boldsymbol{\theta}} f(\boldsymbol{\theta})}{f(\boldsymbol{\theta})}$, the expectation of the gradient of the conditional score w.r.t. $\boldsymbol{\theta}$ satisfies the following expression:

$$\begin{aligned}\mathbb{E}_{\boldsymbol{y} \sim p_{\boldsymbol{\theta}}(\cdot|\boldsymbol{X})} \left[\boldsymbol{s}_{\boldsymbol{Y}|\boldsymbol{X}} (\boldsymbol{\theta}) \right] &= \boldsymbol{\nabla}_{\boldsymbol{\theta}} \mathbb{E}_{\boldsymbol{y} \sim p_{\boldsymbol{\theta}}(\cdot|\boldsymbol{X})} \left[\ln p_{\boldsymbol{\theta}} (\boldsymbol{Y}|\boldsymbol{X}) \right] \\ &\quad - \mathbb{E}_{\boldsymbol{y} \sim p_{\boldsymbol{\theta}}(\cdot|\boldsymbol{X})} \left[\ln p_{\boldsymbol{\theta}} (\boldsymbol{Y}|\boldsymbol{X}) \, \boldsymbol{s}_{\boldsymbol{Y}|\boldsymbol{X}} (\boldsymbol{\theta}) \right]\end{aligned}$$

17 The Leibniz integral rule is granted by the C2 condition. Therefore, the average score satisfies the following expression:

$$\mathbb{E}_{\boldsymbol{y} \sim p_{\boldsymbol{\theta}^*}(\cdot|\boldsymbol{X})} \left[\boldsymbol{s}_{\boldsymbol{Y}|\boldsymbol{X}} (\boldsymbol{\theta}^*) \right] = \boldsymbol{\nabla}_{\boldsymbol{\theta}} \mathbb{E}_{\boldsymbol{y} \sim p_{\boldsymbol{\theta}^*}(\cdot|\boldsymbol{X})} [1] = \mathbf{0}$$

which implies that :

$$\boldsymbol{\nabla}_{\boldsymbol{\theta}^*} \mathbb{E}_{\boldsymbol{y} \sim p_{\boldsymbol{\theta}^*}(\cdot|\boldsymbol{X})} \left[\ln p_{\boldsymbol{\theta}^*} (\boldsymbol{Y}|\boldsymbol{X}) \right] = \mathbb{E}_{\boldsymbol{y} \sim p_{\boldsymbol{\theta}^*}(\cdot|\boldsymbol{X})} \left[\ln p_{\boldsymbol{\theta}^*} (\boldsymbol{Y}|\boldsymbol{X}) \, \boldsymbol{s}_{\boldsymbol{Y}|\boldsymbol{X}} (\boldsymbol{\theta}^*) \right]$$

18 By replacing in the expectation operator by its empirical counterpart in equation [5.32], we obtain:

$$\hat{\mathbb{E}}_{(\boldsymbol{x},\boldsymbol{y}) \sim p_{\boldsymbol{\theta}^*}} \left[\boldsymbol{s}_{\boldsymbol{Y}|\boldsymbol{X}} (\boldsymbol{\theta}^*; \mathcal{D}_{\mathcal{XY}}) \right] = -\sum_{k=1}^{N} \ln p_{\boldsymbol{\theta}} (\boldsymbol{y}_k | \boldsymbol{x}_k) = 0$$

condition $\hat{\boldsymbol{s}}\left(\hat{\boldsymbol{\theta}}_N; \mathcal{D}_{\mathcal{XY}}\right) = \mathbf{0}$ becomes sufficient to minimize the empirical conditional negative log-likelihood.

REMARK 5.8.– $\left(\hat{\boldsymbol{\theta}}_N\right)_{N\in\mathbb{N}}$ represents a unique sequence converging toward $\boldsymbol{\theta}^*$, since $\forall\boldsymbol{\theta} \in O_\varepsilon(\boldsymbol{\theta}^*)$, with $O_\varepsilon(\boldsymbol{\theta}^*)$ being an open neighborhood[19] of $\boldsymbol{\theta}^*$, the probability associated with the set of observations S_N, defined as:

$$S_N := \left\{(\boldsymbol{x}_k, \boldsymbol{y}_k) \in \mathcal{D}_{\mathcal{XY}} : \ln p_{\boldsymbol{\theta}^*}(\boldsymbol{y}_k|\boldsymbol{x}_k) > \sup_{\boldsymbol{\theta}\in\partial O_\varepsilon(\boldsymbol{\theta}^*)} \ln p_{\boldsymbol{\theta}}(\boldsymbol{y}_k|\boldsymbol{x}_k)\right\}$$

converges to 1:

$$\forall\varepsilon > 0 \quad \lim_{N\to+\infty} \mathbb{P}\left[|\mathbb{P}(S_N) - 1| < \varepsilon\right] = 1$$

which means that all the samples, when the data sequence of realization is sufficiently large, most probably belong to S_N, with the maximum log-likelihood over the closure $\bar{O}_\varepsilon(\boldsymbol{\theta}^*)$ corresponding to a point $\boldsymbol{\theta}^* \in \mathring{\Theta}$ (interior of Θ) (Campagne 2022). Because of the regularity condition C2, $p_{\boldsymbol{\theta}}$ is continuous and differentiable over $O_\varepsilon(\boldsymbol{\theta}^*)$. Thanks to Rolle's theorem, $\exists\hat{\boldsymbol{\theta}}_N \in \bar{O}_\varepsilon(\boldsymbol{\theta}^*)$ such that $\boldsymbol{\nabla}_{\boldsymbol{\theta}} \ln p_{\hat{\boldsymbol{\theta}}_N}(\boldsymbol{y}|\boldsymbol{x}) = \mathbf{0}$. This implies that $S_N \subset \tilde{S}_N$, defined as:

$$\tilde{S}_N := \left\{(\boldsymbol{x}_k, \boldsymbol{y}_k) \in \mathcal{D}_{\mathcal{XY}} | \exists\hat{\boldsymbol{\theta}}_N \in \bar{O}_\varepsilon(\boldsymbol{\theta}^*) \text{ such that } \boldsymbol{\nabla}_{\boldsymbol{\theta}} \ln p_{\hat{\boldsymbol{\theta}}_N}(\boldsymbol{y}_k|\boldsymbol{x}_k) = \mathbf{0}\right\}$$

since a priori $\ln p_{\boldsymbol{\theta}^*}(\boldsymbol{y}|\boldsymbol{x}) > \ln p_{\hat{\boldsymbol{\theta}}_N}(\boldsymbol{y}|\boldsymbol{x})$ is not granted. Therefore, $\mathbb{P}(S_N) \leq \mathbb{P}\left(\tilde{S}_N\right) \leq 1$. Due to the convergence in probability of $\mathbb{P}(S_N)$ to 1, then $\mathbb{P}\left(\tilde{S}_N\right)$ converges in probability to 1, which means that for $\forall\varepsilon > 0$, there exists an open neighborhood $O_\varepsilon(\boldsymbol{\theta}^*)$ to which $\hat{\boldsymbol{\theta}}_N$ most probably belongs, converging most probably toward $\boldsymbol{\theta}^*$, by definition.

---

19 $\varepsilon > 0$

Last but not least, adopting a weakly unbiased estimator $\hat{\boldsymbol{\theta}}_N$ and considering it as a continuous random vector, the probability of the event $\hat{\boldsymbol{\theta}}_N = \boldsymbol{\theta}^*$ is null (because the singleton $\{\boldsymbol{\theta}^*\}$ has null Lebesgue measure).

5.2.3. *The Fisher information: geometric interpretation*

In his seminal work, Fisher formulated the concept of *information* associated with the parameters $\boldsymbol{\theta}$ and quantified as the covariance of $s_{\boldsymbol{Y}|\boldsymbol{X}}(\boldsymbol{\theta})$:

$$
\begin{aligned}
\mathbb{I}^F_{\boldsymbol{Y}|\boldsymbol{X}}(\boldsymbol{\theta}^*) &= \mathbb{E}_{(\boldsymbol{x},\boldsymbol{y})\sim p_{\boldsymbol{\theta}^*}}\left[s_{\boldsymbol{Y}|\boldsymbol{X}}(\boldsymbol{\theta}^*) \otimes s_{\boldsymbol{Y}|\boldsymbol{X}}(\boldsymbol{\theta}^*)\right] \\
&= \mathbb{C}_{(\boldsymbol{x},\boldsymbol{y})\sim p_{\boldsymbol{\theta}^*}}\left[s_{\boldsymbol{Y}|\boldsymbol{X}}(\boldsymbol{\theta}^*)\right] \geq 0
\end{aligned} \tag{5.33}
$$

$\mathbb{I}^F_{\boldsymbol{Y}|\boldsymbol{X}}(\boldsymbol{\theta}^*)$ in equation [5.33] is called the conditional *Fisher Information Matrix* (FIM), and it is a $d_\Theta \times d_\Theta$ positive semidefinite matrix. If $\mathbb{I}^F_{\boldsymbol{Y}|\boldsymbol{X}}(\boldsymbol{\theta}^*)$ is positive definite, then it defines a Riemann metric on the d_Θ-dimensional parameter space. For i.i.d. samples (which is equivalent to adopting the empirical measure), the Fisher information goes under the name of *observed Fisher information* and it reads:

$$
\begin{aligned}
\hat{\mathbb{I}}^F_{\boldsymbol{Y}|\boldsymbol{X}}(\boldsymbol{\theta}^*; \mathcal{D}_{\mathcal{XY}}) &= \sum_{k=1}^{N}\left[\boldsymbol{\nabla}_{\boldsymbol{\theta}} \ln p_{\boldsymbol{\theta}^*}(\boldsymbol{y}_k|\boldsymbol{x}_k) \otimes \boldsymbol{\nabla}_{\boldsymbol{\theta}} \ln p_{\boldsymbol{\theta}^*}(\boldsymbol{y}_k|\boldsymbol{x}_k)\right] \\
&= N \cdot \mathbb{I}^F_{\boldsymbol{y}|\boldsymbol{x}}(\boldsymbol{\theta}^*)
\end{aligned} \tag{5.34}
$$

because the score function is centered (see equation [5.32]) (Billingsley 1995; Cournède 2020; Campagne 2022). $\mathbb{I}^F_{\boldsymbol{y}|\boldsymbol{x}}(\boldsymbol{\theta}^*)$ is the observed Fisher information associated with any arbitrary sample (since all samples are i.i.d.). equation [5.34] is another way to express the fact that the variance of a sum of i.i.d. variable is the sum of the variances. A set of i.i.d. random variables, with high conditional Fisher information, has a score with large variance, that is, it is *informative*. The "narrower" (in average on $\mathcal{D}_{\mathcal{XY}}$) is the $NLL_{\boldsymbol{Y}|\boldsymbol{X}}$ minimum, the more informative is $\boldsymbol{\theta}^*$ and therefore the higher is the score covariance (Campagne 2022; Cournède 2020, the largest the span of data around the minimum). Moreover, under C1, C2, C3 and C4 regularity conditions and for $\boldsymbol{\nabla}_{\boldsymbol{\theta}} p_{\boldsymbol{\theta}^*} \in \left(\mathcal{C}^1(\Theta)\right)^{d_\Theta}$, the conditional Fisher's information can be also expressed by the second Bartlett identity, which reads:

$$\mathbb{I}^F_{\boldsymbol{Y}|\boldsymbol{X}}(\boldsymbol{\theta}^*) = \mathbb{E}_{(\boldsymbol{x},\boldsymbol{y})\sim p_{\boldsymbol{\theta}^*}}\left[\mathbb{H}_{NLL_{\boldsymbol{Y}|\boldsymbol{X}}}(\boldsymbol{\theta}^*)\right] \qquad [5.35]$$

with $\mathbb{H}_{NLL_{\boldsymbol{Y}|\boldsymbol{X}}}$ being the Hessian matrix of conditional negative log-likelihood (see Cournède 2020, for the proof). The Fisher information may be seen as the expected curvature $\kappa\left(\hat{\boldsymbol{\theta}}_N\right)$ of the $NLL_{\boldsymbol{Y}|\boldsymbol{X}}$, for $\boldsymbol{\theta}$ infinitesimally close to the unbiased estimator $\hat{\boldsymbol{\theta}}_N$. Figure 5.5 shows a geometrical interpretation of the FIM for regular statistical model (i.e. positive definite, with positive definite Hessian of $NLL_{\boldsymbol{Y}|\boldsymbol{X}}$): the inverse of the absolute curvature $\kappa\left(\hat{\boldsymbol{\theta}}_N\right)$ of the $NLL_{\boldsymbol{Y}|\boldsymbol{X}}$ is the radius $r(\boldsymbol{\theta}) = \frac{1}{\kappa\left(\hat{\boldsymbol{\theta}}_N\right)}$ of the osculating circle at the MLE $\hat{\boldsymbol{\theta}}_N$ (Nielsen 2020). The larger is the spectral radius of the conditional Fisher information, the larger the absolute curvature of $\kappa\left(\hat{\boldsymbol{\theta}}_N\right)$ (for i.i.d., the larger is observed information of any single sample, $\mathbb{I}^F_{\boldsymbol{y}_1|\boldsymbol{x}_1}(\boldsymbol{\theta}^*)$, because of equation [5.34]), the sharper is the peak at $\hat{\boldsymbol{\theta}}_N$ (because of small curvature radius) and the more precise the MLE.

5.2.4. *The Fisher information: statistical interpretation*

REMARK 5.9.– The FIM of an MVN $\mathcal{N}\left(\boldsymbol{\mu}\left(\boldsymbol{\theta}_{\boldsymbol{\mu}}\right), \boldsymbol{\Sigma}\left(\boldsymbol{\theta}_{\boldsymbol{\Sigma}}\right)\right)$ reads:

$$\mathbb{I}^F(\boldsymbol{\theta}) = \left(\boldsymbol{\nabla}_{\boldsymbol{\theta}_{\boldsymbol{\mu}}} \otimes \boldsymbol{\mu}\right).\boldsymbol{\Sigma}^{-1}.\left(\boldsymbol{\mu} \otimes \boldsymbol{\nabla}_{\boldsymbol{\theta}_{\boldsymbol{\mu}}}\right)$$
$$+\frac{1}{2}\sum_{i,j=1}^{d_\Theta} \mathrm{Tr}\left(\boldsymbol{\Sigma}^{-1}\frac{\partial \boldsymbol{\Sigma}}{\partial \theta_{\boldsymbol{\Sigma},i}}\boldsymbol{\Sigma}^{-1}\frac{\partial \boldsymbol{\Sigma}}{\partial \theta_{\boldsymbol{\Sigma},j}}\right) e_i \otimes e_j \qquad [5.36]$$

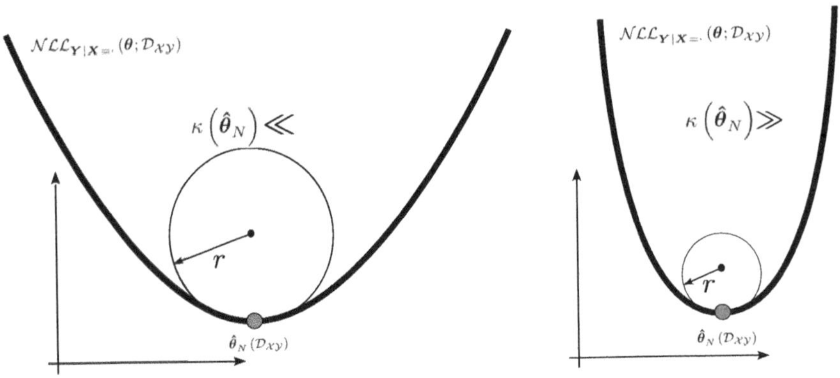

Figure 5.5. *Geometrical representation of the conditional $NLL_{\mathbf{Y}|\mathbf{X}}$ minimization problem. (a) Small FI/curvature (flat and spread-out peak) → large variance (small accuracy); (b) large FI/curvature (sharp peak) → small variance (high accuracy). Inspired by F. Nielsen (https://franknielsen.github.io/SlidesVideo/PrintIntroductionInformationGeometry-FrankNielsen.pdf)*

If $\boldsymbol{\mu} = \boldsymbol{\theta}_{\boldsymbol{\mu}}$ and $\boldsymbol{\Sigma} = \boldsymbol{\theta}_{\boldsymbol{\Sigma}}$, then the FIM reads:

$$\mathbb{I}^F(\boldsymbol{\theta}) = \boldsymbol{\Sigma}^{-1} + \frac{1}{2}\mathrm{Tr}\left(\boldsymbol{\Sigma}^{-2}\right)\mathbb{I}$$

These latter findings show that the FIM of an MVN $\mathcal{N}(\boldsymbol{\mu}, \boldsymbol{\Sigma})$ does not depend on the bias $\boldsymbol{\mu}$ and it increases for larger eigenvalues of the *precision matrix* $\boldsymbol{\Sigma}^{-1}$, that is, for the smaller spectral radius of the covariance matrix $\boldsymbol{\Sigma}$. In an empirical framework, the FIM is high when the estimator $\hat{\boldsymbol{\theta}}_N$ is predicted with high accuracy and lower variance (corresponding to sharper peak, according to the geometric interpretation provided in the previous section). In the following, the asymptotic behavior of $\hat{\boldsymbol{\theta}}_N$ is studied to assess how close it is to the true value $\boldsymbol{\theta}^*$.

Assuming, from now on, that the MLE is square-integrable, that is, $\mathbb{E}_{(\boldsymbol{x},\boldsymbol{y})\sim p_{\boldsymbol{\theta}^*}}\left[\|\hat{\boldsymbol{\theta}}_N(\mathcal{D}_{\mathcal{XY}})\|\right] < +\infty$, to assess how close $\hat{\boldsymbol{\theta}}_N(\mathcal{D}_{\mathcal{XY}})$ is to the true parameter set $\boldsymbol{\theta}^*$, the *quadratic risk* $R(\hat{\boldsymbol{\theta}}_N, \boldsymbol{\theta}^*)$ is adopted[20], defined as (Cournède 2020)[21]:

$$R(\hat{\boldsymbol{\theta}}_N, \boldsymbol{\theta}^*) = \mathbb{E}_{(\boldsymbol{x},\boldsymbol{y})\sim p_{\boldsymbol{\theta}^*}} \left[\|\hat{\boldsymbol{\theta}}_N (\mathcal{D}_{\mathcal{XY}}) - \boldsymbol{\theta}^* \|^2 \right] = \\ = \mathrm{Tr} \left(\mathbb{C}_{(\boldsymbol{x},\boldsymbol{y})\sim p_{\boldsymbol{\theta}^*}} \left[\hat{\boldsymbol{\theta}}_N (\mathcal{D}_{\mathcal{XY}}) \right] \right) + \| \boldsymbol{b}_{\boldsymbol{\theta}^*} \left(\hat{\boldsymbol{\theta}}_N (\mathcal{D}_{\mathcal{XY}}) \right) \|^2 \tag{5.37}$$

with $\boldsymbol{b}_{\boldsymbol{\theta}^*} \left(\hat{\boldsymbol{\theta}}_N (\mathcal{D}_{\mathcal{XY}}) \right) = \mathbb{E}_{(\boldsymbol{x},\boldsymbol{y})\sim p_{\boldsymbol{\theta}^*}} \left[\hat{\boldsymbol{\theta}}_N (\mathcal{D}_{\mathcal{XY}}) \right] - \boldsymbol{\theta}^*$ the so-called *bias* and $\mathbb{C}_{(\boldsymbol{x},\boldsymbol{y})\sim p_{\boldsymbol{\theta}^*}} \left[\hat{\boldsymbol{\theta}}_N (\mathcal{D}_{\mathcal{XY}}) \right]$ the covariance matrix of $\hat{\boldsymbol{\theta}}_N$. To reduce the quadratic risk in equation [5.37] one must choose the best-unbiased estimator $\hat{\boldsymbol{\theta}}_N$, for which $\boldsymbol{b}_{\boldsymbol{\theta}^*} \left(\hat{\boldsymbol{\theta}}_N (\mathcal{D}_{\mathcal{XY}}) \right) = \mathbf{0}$ (*i.e.*, such that the average estimator is equal to the true parameter set, $\mathbb{E}_{(\boldsymbol{x},\boldsymbol{y})\sim p_{\boldsymbol{\theta}^*}} \left[\hat{\boldsymbol{\theta}}_N (\mathcal{D}_{\mathcal{XY}}) \right] = \boldsymbol{\theta}^*$). Moreover, the best choice of $\hat{\boldsymbol{\theta}}_N$ must correspond to the lowest variance, *i.e.*, such that for all $\hat{\boldsymbol{\theta}}'_N$ it holds that $\mathbb{C}_{(\boldsymbol{x},\boldsymbol{y})\sim p_{\boldsymbol{\theta}}} \left[\hat{\boldsymbol{\theta}}_N (\mathcal{D}_{\mathcal{XY}}) \right] \preccurlyeq \mathbb{C}_{(\boldsymbol{x},\boldsymbol{y})\sim p_{\boldsymbol{\theta}}} \left[\hat{\boldsymbol{\theta}}'_N (\mathcal{D}_{\mathcal{XY}}) \right]$, $\forall \boldsymbol{\theta} \in \Theta$ and it exists $\boldsymbol{\theta}'$ such that $\mathbb{C}_{(\boldsymbol{x},\boldsymbol{y})\sim p_{\boldsymbol{\theta}'}} \left[\hat{\boldsymbol{\theta}}_N (\mathcal{D}_{\mathcal{XY}}) \right] \prec \mathbb{C}_{(\boldsymbol{x},\boldsymbol{y})\sim p_{\boldsymbol{\theta}'}} \left[\hat{\boldsymbol{\theta}}'_N (\mathcal{D}_{\mathcal{XY}}) \right]$[22]. For regular statistical models, the

---

20 If $\forall \boldsymbol{\theta} \in \Theta$, $R\left(\hat{\boldsymbol{\theta}}_N, \boldsymbol{\theta}\right) \leq R\left(\hat{\boldsymbol{\theta}}'_N, \boldsymbol{\theta}\right)$ and $\exists \boldsymbol{\theta}' \in \Theta$ such that $R\left(\hat{\boldsymbol{\theta}}_N, \boldsymbol{\theta}'\right) < R\left(\hat{\boldsymbol{\theta}}'_N, \boldsymbol{\theta}'\right)$, then $\hat{\boldsymbol{\theta}}'_N$ is inadmissible and $\hat{\boldsymbol{\theta}}_N$ is a better estimator, admissible if it is the best.

21 Note that by calling $\boldsymbol{b}_{\boldsymbol{\theta}^*} \left(\hat{\boldsymbol{\theta}}_N (\mathcal{D}_{\mathcal{XY}}) \right) = \mathbb{E}_{(\boldsymbol{x},\boldsymbol{y})\sim p_{\boldsymbol{\theta}^*}} \left[\hat{\boldsymbol{\theta}}_N (\mathcal{D}_{\mathcal{XY}}) - \boldsymbol{\theta}^* \right]$, it holds that:

$$\mathbb{E}_{(\boldsymbol{x},\boldsymbol{y})\sim p_{\boldsymbol{\theta}^*}} \left[\|\hat{\boldsymbol{\theta}}_N - \boldsymbol{\theta}^* - \boldsymbol{b}_{\boldsymbol{\theta}^*} \|^2 \right] = R(\hat{\boldsymbol{\theta}}_N, \boldsymbol{\theta}^*) - \|\boldsymbol{b}_{\boldsymbol{\theta}^*}\|^2$$

and

$$\mathbb{E}_{(\boldsymbol{x},\boldsymbol{y})\sim p_{\boldsymbol{\theta}^*}} \left[\left(\hat{\boldsymbol{\theta}}_N (\mathcal{D}_{\mathcal{XY}}) - \boldsymbol{\theta}^* - \boldsymbol{b}_{\boldsymbol{\theta}^*} \right) \otimes \left(\hat{\boldsymbol{\theta}}_N (\mathcal{D}_{\mathcal{XY}}) - \boldsymbol{\theta}^* - \boldsymbol{b}_{\boldsymbol{\theta}^*} \right) \right] \\ = \mathbb{C}_{(\boldsymbol{x},\boldsymbol{y})\sim p_{\boldsymbol{\theta}^*}} \left[\hat{\boldsymbol{\theta}}_N (\mathcal{D}_{\mathcal{XY}}) - \boldsymbol{\theta}^* \right]$$

22 $\mathbb{A} \preccurlyeq \mathbb{B}$ if and only if $\mathbb{B} - \mathbb{A}$ is positive semidefinite; $\mathbb{A} \prec \mathbb{B}$ if and only if $\mathbb{B} - \mathbb{A}$ is positive definite.

covariance matrix of an unbiased estimator $\hat{\boldsymbol{\theta}}_N$ is represented by the inverse conditional FIM $\hat{\mathbb{I}}^{F}_{\boldsymbol{Y}|\boldsymbol{X}}{}^{-1}$, commonly known as Cramér–Rao lower bound (CRB):

THEOREM 5.2 (**Cramér–Rao bound**).– Provided a regular statistical model $\mathcal{H}_\Theta := \left\{P_{\boldsymbol{\theta}}, \boldsymbol{\theta} \in \Theta \subset \mathbb{R}^{d_\Theta}\right\} \ni p_{\boldsymbol{\theta}^*}$ (that at least satisfies C1 and C2 conditions) and a square-integrable estimator of the true parameter set $\boldsymbol{\theta}^*$, namely $\hat{\boldsymbol{\theta}}_N\left(\mathcal{D}_{\mathcal{X}\mathcal{Y}}\right)$, with invertible observed conditional FIM and whose expectation reads:

$$\bar{\boldsymbol{\theta}}_N = \mathbb{E}_{(\boldsymbol{x},\boldsymbol{y})\sim p_{\boldsymbol{\theta}^*}}\left[\hat{\boldsymbol{\theta}}_N\left(\mathcal{D}_{\mathcal{X}\mathcal{Y}}\right)\right]$$

if $\boldsymbol{\nabla}_{\boldsymbol{\theta}} \otimes \bar{\boldsymbol{\theta}}_N = \mathbb{E}_{(\boldsymbol{x},\boldsymbol{y})\sim p_{\boldsymbol{\theta}^*}}\left[\boldsymbol{\nabla}_{\boldsymbol{\theta}} \otimes \hat{\boldsymbol{\theta}}_N\left(\mathcal{D}_{\mathcal{X}\mathcal{Y}}\right)\right]$, then the variance of the estimator is bounded by below as follows:

$$\mathbb{V}_{(\boldsymbol{x},\boldsymbol{y})\sim p_{\boldsymbol{\theta}^*}}\left[\hat{\boldsymbol{\theta}}_N\right] \geq \frac{\|\boldsymbol{\nabla}_{\boldsymbol{\theta}} \otimes \bar{\boldsymbol{\theta}}_N\|^2}{N \cdot \mathbb{I}^F_{\boldsymbol{y}|\boldsymbol{x}}(\boldsymbol{\theta}^*)} \quad [5.38]$$

$\hat{\boldsymbol{\theta}}_N\left(\mathcal{D}_{\mathcal{X}\mathcal{Y}}\right)$ is unbiased if:

$$\mathbb{V}_{(\boldsymbol{x},\boldsymbol{y})\sim p_{\boldsymbol{\theta}^*}}\left[\hat{\boldsymbol{\theta}}_N\right] \geq \frac{1}{N \cdot \mathbb{I}^F_{\boldsymbol{y}|\boldsymbol{x}}(\boldsymbol{\theta}^*)} \quad [5.39]$$

The following theorem adds that, provided an extra regularity condition of the statistical model and thanks to the CRB, the MLE estimator converges in standard normal probability to $\boldsymbol{\theta}^*$.

THEOREM 5.3 (**MLE convergence in standard normal probability**; Cournède 2020; Campagne 2022).– Provided a statistical model $\mathcal{H}_\Theta := \left\{P_{\boldsymbol{\theta}}, \boldsymbol{\theta} \in \Theta \subset \mathbb{R}^{d_\Theta}\right\}$ with regularity conditions C1, C2, C3, C4 and an i.i.d. dataset $\mathcal{D}_{\mathcal{X}\mathcal{Y}}$, generated by a parameter $\boldsymbol{\theta}^* \in \mathring{\Theta}$, with and positive definite FIM in a neighborhood of $\boldsymbol{\theta}^*$ and the conditional MLE $\hat{\boldsymbol{\theta}}_N = \underset{\boldsymbol{\theta}\in\Theta}{\arg\min}\, NLL_{\boldsymbol{Y}|\boldsymbol{X}}\left(\boldsymbol{\theta}; \mathcal{D}_{\mathcal{X}\mathcal{Y}}\right)$, if a further regularity condition C5 of $\mathcal{H}_\Theta$ holds, namely:

$$\begin{aligned}&\left\|\boldsymbol{\nabla}_{\boldsymbol{\theta}}\otimes\boldsymbol{\nabla}_{\boldsymbol{\theta}}\otimes\boldsymbol{s}\left(\boldsymbol{Y}|\boldsymbol{X}\right)\right\|<M\left(\boldsymbol{X},\boldsymbol{Y}\right)\\&\forall\boldsymbol{\theta}\in\Theta,\mathbb{E}_{(\boldsymbol{x},\boldsymbol{y})\in p_{\boldsymbol{\theta}}}\left[M\right]<+\infty\end{aligned} \quad [5.40]$$

then all MLE sequences $\hat{\boldsymbol{\theta}}_N$ converge in standard normal probability to $\boldsymbol{\theta}^*$:

$$\forall\varepsilon>0 \quad \lim_{N\to+\infty}\mathbb{P}\left[\left|p_{\sqrt{N}\left(\hat{\boldsymbol{\theta}}_N-\boldsymbol{\theta}^*\right)}-\mathcal{N}\left(\mathbf{0},\mathbb{I}_F^{-1}\left(\boldsymbol{\theta}^*;\mathcal{D}_{\mathcal{X}}\right)\right)\right|<\varepsilon\right]=1 \quad [5.41]$$

Therefore, the MLE is an asymptotically unbiased estimator and thanks to equation [5.39], one can infer that the higher the observed FIM, the lower will be its variance close to the minimum of the negative log-likelihood (sharper peak).

REMARK 5.10.– Theorem 5.3 establishes the confidence intervals of the MLE algorithm, providing the possibility of assessing the probability (standard normal) that the MLE approaches the "true" probability distribution $p_{\boldsymbol{\theta}^*}\left(\boldsymbol{x}\right)$ (Campagne 2022).

REMARK 5.11.– Provided the CRB for an unbiased estimator, its *efficiency* can be measured as the ratio between the variance lower bound (expressed by the CRB) and the variance of the estimator, namely:

$$\eta\left(\hat{\boldsymbol{\theta}}_N\right)=\frac{1}{N\cdot\mathbb{I}^F_{\boldsymbol{y}|\boldsymbol{x}}\left(\boldsymbol{\theta}^*\right)\cdot\mathbb{V}_{(\boldsymbol{x},\boldsymbol{y})\sim p_{\boldsymbol{\theta}^*}}\left(\hat{\boldsymbol{\theta}}_N\right)}\leq 1 \quad [5.42]$$

Therefore, because of the CRB, the MLE is an asymptotically "optimum" estimator, that is, its efficiency $\eta\left(\hat{\boldsymbol{\theta}}_N\right)$ (defined in equation [5.42]) converges in probability to 1:

$$\forall\varepsilon>0 \quad \lim_{N\to\infty}\mathbb{P}\left[\left|\eta\left(\hat{\boldsymbol{\theta}}_N\right)-1\right|<\varepsilon\right]=1 \quad [5.43]$$

REMARK 5.12.– The CRB states that the higher the information of an estimator $\hat{\boldsymbol{\theta}}_N$, the higher the variance of the score and, because of the CRB, the lower its variance (the narrower is the NLL minimum). The FIM and the CRB steer the convergence rate of any algorithm that attempts to maximize the log-likelihood of any statistical model adopted to infer from the available dataset. However, from a practical standpoint, the FIM matrix (dense) can easily reach an unbearable computational cost, which is why, in practice, it is rarely explicitly adopted. First-order gradient descent methods are adopted instead (see section 5.3).

A NN is a particular statistical model $\boldsymbol{h}_{\boldsymbol{\theta}} \in \mathcal{H}_\Theta$ that depends on a set of weights and biases $\boldsymbol{\theta} \in \Theta$. The core of statistical learning with NN is represented by the *empirical loss/risk minimization* problem:

PROBLEM 5.1 **(Empirical loss minimization** $(\mathcal{P})$**).**– Find $\boldsymbol{h}_{\boldsymbol{\theta}} : \mathcal{X} \to \mathcal{Y}, \boldsymbol{\theta} \in \Theta$, such that:

$$\hat{\boldsymbol{\theta}}_N (\mathcal{D}_{\mathcal{X}\mathcal{Y}}) = \underset{\boldsymbol{\theta}\in\Theta}{\arg\min}\; \mathcal{L}(\boldsymbol{h}_{\boldsymbol{\theta}}; \mathcal{D}_{\mathcal{X}\mathcal{Y}})$$

where $\mathcal{L}$ reads:

$$\mathcal{L}\left(\boldsymbol{h}_{\boldsymbol{\theta}}; \mathcal{D}_{\mathcal{X}\mathcal{Y}}\right) = \frac{1}{N} NLL_{\mathbf{Y}|\mathbf{X}} \left(\boldsymbol{\theta}; \mathcal{D}_{\mathcal{X}\mathcal{Y}}\right) + \Omega(\boldsymbol{\theta}) \qquad [5.44]$$

In general, there might be no solution, in the sense that $\underset{\boldsymbol{\theta}\in\Theta}{\arg\min}\mathcal{L}(\boldsymbol{\theta}) =$ or that $\underset{\boldsymbol{\theta}\in\Theta}{\min}\mathcal{L} = m$ is finite, but for $||\boldsymbol{\theta}|| \to +\infty$. If at least on minimizer $\hat{\boldsymbol{\theta}}_N$ exists and $\hat{\boldsymbol{\theta}}_N \in \Theta$, Θ must be a compact subset of $\mathbb{R}^{d_\Theta}$ and $\mathcal{L}$ lower semi-continuous on it (Peyré 2020). This topic is discussed in depth in section 5.3.

REMARK 5.13.– For NN, the Fisher approach can fail because $\boldsymbol{\theta}^*$ is not unique, since the loss function $\mathcal{L}$ is not convex (Campagne 2022). Moreover, in practice, the dimension of the parameter space $d_\Theta \gg d_X$ implies that the problem is over-parameterized, which makes it difficult to find unbiased estimators (Campagne 2022).

REMARK 5.14.– $\Omega(\boldsymbol{\theta})$ is a *regularization* term that prevents *overfitting*, an unpleasant phenomenon that occurs when the model $\boldsymbol{h}_{\boldsymbol{\theta}}$ has enough parameters to reach the MDI with the available data $\mathcal{D}_{\mathcal{XY}}$, that is, $\boldsymbol{h}_{\boldsymbol{\theta}}$ minimizes the empirical Kullback–Leibler divergence between true distribution and the statistical model it parameterizes (see equation [5.28]). However, this fortunate coincidence corresponds to considering that the true probability is solely supported by the discrete set of data, leaving no alternative for novel data $(\boldsymbol{x}_{NEW}, \boldsymbol{y}_{NEW})$, whose probability $p_{\hat{\boldsymbol{\theta}}_N}(\boldsymbol{x}_{NEW}, \boldsymbol{y}_{NEW}) = 0$. $\boldsymbol{h}_{\boldsymbol{\theta}}$ cannot generalize in this case, unless regularization is added.

REMARK 5.15.– Adopting the Bayesian framework for which the empirical statistical model $p_{\boldsymbol{\theta}}(\boldsymbol{Y}|\boldsymbol{X}, \mathcal{D}_{\mathcal{XY}}) = p(\boldsymbol{Y}|\boldsymbol{X}, \mathcal{D}_{\mathcal{XY}}, \boldsymbol{\theta})$, then:

$$\ln p_{\boldsymbol{\theta}}(\boldsymbol{Y}|\boldsymbol{X}, \mathcal{D}_{\mathcal{XY}}) = \ln p_{\boldsymbol{\theta}}(\boldsymbol{Y}|\boldsymbol{X}, \mathcal{D}_{\mathcal{XY}}) + \ln p(\boldsymbol{\theta})$$

$\Omega(\boldsymbol{\theta})$ is usually considered as $-\ln p(\boldsymbol{\theta})$, leading to a maximum a posterior (MAP) estimation. The arbitrary choice of the prior $p(\boldsymbol{\theta})$ leads to different regularization, such as if $p(\boldsymbol{\theta}) = \mathcal{N}\left(\mathbf{0}, \sqrt{\frac{2}{\lambda}\mathbb{I}}\right)$ $(\lambda \geq 0)$, with $\Omega(\boldsymbol{\theta}) = \lambda\|\boldsymbol{\theta}\|_2$+const. The choice of a Laplace distribution, instead, leads to a ℓ_1 regularization term (see equation [2.20] for further details).

REMARK 5.16.– In a regression framework, the usual choice of $\mathcal{L}$ is the MSE $\mathcal{L}(\boldsymbol{\theta}) = \frac{1}{N}\sum_{k=1}^{N}\|\boldsymbol{y}_k - \boldsymbol{h}_{\boldsymbol{\theta}}(\boldsymbol{x}_k)\|$ (see equation [2.7]). This comes from the choice of a homoscedastic MVN as a likelihood function (see equation [5.27]). For multiclass classification purposes, the chosen likelihood is usually:

$$p_{\boldsymbol{\theta}}(\mathcal{D}_{\mathcal{XY}}) = Cat\left(y|\boldsymbol{h}_{\boldsymbol{\theta}}\right) = \prod_{k=1}^{N}\prod_{c=1}^{N_C} h_{\boldsymbol{\theta},c}^{1_{y_k=c}}(\boldsymbol{x}_k) \qquad [5.45]$$

with $y_k \in \mathcal{Y} = \{1, \ldots, N_C\}$. In this case, $h_{\boldsymbol{\theta},c}(x_k) = \mathbb{P}(\{y_k = c\} | x_k, \boldsymbol{\theta})$, so it is required that $0 \leq h_{\boldsymbol{\theta},c} \leq 1$ and $\prod_{c=1}^{N_C} h_{\boldsymbol{\theta},c}(x_k) = 1, \forall x_k \in \mathcal{X}$ (see section 4.2.4.2).

EXAMPLE 5.3 (**Empirical loss in** `PyTorch`).– In `PyTorch`, several different precoded loss functions are available and provided by the library `torch.nn`. The list is available at: https://pytorch.org/docs/stable/nn.html#loss-functions.

```
import torch

# Define loss function (MLE for regression)
loss_fn1 = torch.nn.MSELoss(reduction='mean')

loss_fn2 = torch.nn.NLLLoss(reduction='sum')

loss_fn3 = torch.nn.BCELoss(weight=None, reduction='none')
```

Different types of *reductions* are available: none, empirical mean (`reduction='mean'`) or simple sum (`reduction='sum'`). For some losses, such as the Binary Cross Entropy (`torch.nn.BCELoss`) adopted for binary classification (and corresponding to the $NLL_{\boldsymbol{Y}|\boldsymbol{X}}(\boldsymbol{\theta}; \mathcal{D}_{\mathcal{XY}})$ for the classification likelihood in equation [5.45], with two classes only, $N_C = 2$, see section 4.2.2), `PyTorch` offers the choice of weighting the loss computed for each sample $(\boldsymbol{x}_k, \boldsymbol{y}_k)$ differently.

5.2.5. *The principle of maximum entropy (MaxEnt)*

As shown in section 5.2.1, the easiest strategy to calibrate a statistical model is the *substitution method*, which employs sufficient number of conditional statistics $S_k(\boldsymbol{X}, \boldsymbol{Y})$ computed on the dataset $\mathcal{D}_{\mathcal{XY}}$, for the joint distribution, to reduce the discrepancy between the characteristics $\mathcal{G}$ of true unknown probability distribution and the statistical model. Those characteristics or "features" can be generically defined as (Murphy 2022):

$$G_k(\mathcal{D}_{\mathcal{XY}}) = \int_{\mathcal{X}} \int_{\mathcal{Y}} g_k(\boldsymbol{x}, \boldsymbol{y}) \hat{P}(\boldsymbol{dx}, \boldsymbol{dy}) \quad [5.46]$$

For instance, the empirical conditional moments $\hat{\boldsymbol{m}}^{(p)}_{\boldsymbol{Y}|\boldsymbol{X}}(\mathcal{D}_{\mathcal{XY}})$ serve as statistics and features of the probability distribution. To this end, the *law of total expectation* states that the conditional empirical average can be obtained according to the following expression (Murphy 2022):

$$\hat{\boldsymbol{m}}^{(1)}_{\boldsymbol{Y}|\boldsymbol{X}}(\mathcal{D}_{\mathcal{XY}}) = \hat{\mathbb{E}}_{\mathcal{Y}}[\boldsymbol{Y}] = \hat{\mathbb{E}}_{\mathcal{X}}\left[\hat{\mathbb{E}}_{\mathcal{Y}}[\boldsymbol{Y}|\boldsymbol{X}]\right]$$

Again, the *law of total variance* states that the conditional covariance (or the second-order moment $\hat{\boldsymbol{m}}^{(2)}_{\boldsymbol{Y}|\boldsymbol{X}}$):

$$\hat{\mathbb{C}}_{\boldsymbol{Y}|\boldsymbol{X}}(\mathcal{D}_{\mathcal{XY}}) = \hat{\mathbb{E}}_{\mathcal{X}}\left[\hat{\mathbb{C}}_{\boldsymbol{Y}}[\boldsymbol{Y}|\boldsymbol{X}]\right] + \hat{\mathbb{C}}_{\boldsymbol{X}}\left[\hat{\mathbb{E}}_{\mathcal{Y}}[\boldsymbol{Y}|\boldsymbol{X}]\right]$$

However, provided that true probability distribution can be parameterized with a parameter $\boldsymbol{\theta}^*$, what statistical model $\mathcal{H}_\Theta$ shall be chosen? Following the *MDI*, the idea is to minimize the empirical Kullback–Leibler divergence between true and model probability distribution. The Fisher approach described above proposes to minimize the conditional $NLL_{\boldsymbol{Y}|\boldsymbol{X}}$, corresponding to the second term of the sum on the right-hand side of the following decomposition of the Kullback–Leibler divergence (reported in its continuous form hereafter, for the sake of clarity, as in equation [5.29]):

$$\begin{aligned}&\mathbb{D}_{KL}\left[P(\boldsymbol{Y}|\boldsymbol{X}) : P_{\boldsymbol{\theta}}(\boldsymbol{Y}|\boldsymbol{X})\right]\\ &= -H_d(\boldsymbol{Y}|\boldsymbol{X}) - \mathbb{E}_{(\boldsymbol{x},\boldsymbol{y})\sim P}\left[\ln p_{\boldsymbol{\theta}}(\boldsymbol{Y}|\boldsymbol{X})\right]\end{aligned} \quad [5.47]$$

The conditional differential entropy $H_d(\boldsymbol{Y}|\boldsymbol{X})$ was so far considered as a constant, independent on the parameters $\boldsymbol{\theta}$, which are instead tuned to minimize the second term $-\mathbb{E}_{(\boldsymbol{x},\boldsymbol{y})\sim P}\left[\ln p_{\boldsymbol{\theta}}(\boldsymbol{Y}|\boldsymbol{X})\right]$. However, this strategy overlooks the pre-emptive choice of $\mathcal{H}_\Theta$. In particular, the best choice would be the one that minimizes the term $-H_d(\boldsymbol{Y}|\boldsymbol{X})$. The conditional entropy $H_d(\boldsymbol{Y}|\boldsymbol{X})$ defined as:

$$H_d(\boldsymbol{Y}|\boldsymbol{X}) = -\mathbb{E}_{\boldsymbol{x},\boldsymbol{y}\sim p_{\boldsymbol{\theta}^*}}\left[\ln p_{\boldsymbol{\theta}^*}(\boldsymbol{Y}|\boldsymbol{X})\right] \qquad [5.48]$$

can be interpreted as a measure of uncertainty/surprise/lack of predictability, associated with a random variable $\boldsymbol{Y}$ provided the value of the other random variable $\boldsymbol{X}$ (Murphy 2022). The discrete distribution with maximum entropy is the uniform distribution. According to (Jaynes 1957), the true probability distribution is associated with the maximum of the conditional Shannon's entropy, that is, the one whose entropy is closer to the uniform discrete one on the finite dataset. Moreover, the following theorem, by Gibbs and Boltzmann, states that the statistical model that maximizes the conditional entropy is the exponential family.

THEOREM 5.4 (**Gibbs–Boltzmann theorem**).– Provided a family of probability distributions $q \in \mathcal{H}$, the solution to the following constrained optimization problem:

$$\begin{cases} \max\limits_{q\in\mathcal{H}} H_d(q) = \max\limits_{q\in\mathcal{H}}\left(-\int_{\mathcal{X}}\int_{\mathcal{Y}} \ln q\left(\boldsymbol{y}|\boldsymbol{x}\right)\cdot q\left(\boldsymbol{x},\boldsymbol{y}\right)\mu(\boldsymbol{dx},\boldsymbol{dy})\right) \\ \text{with } S_k\left(q\right) = \mu_{f_k} - \int_{\mathcal{X}}\int_{\mathcal{X}} f_k\left(\boldsymbol{x},\boldsymbol{y}\right)\cdot q\left(\boldsymbol{x},\boldsymbol{y}\right)\mu(\boldsymbol{dx},\boldsymbol{dy}) = 0, \\ \quad 1 \leq k \leq K \end{cases} \qquad [5.49]$$

if it exists it is defined as $p_{\boldsymbol{\theta}^*} \in \mathcal{H}$, which reads:

$$\begin{aligned} p_{\boldsymbol{\theta}^*}\left(\boldsymbol{y}|\boldsymbol{x}\right) &= \operatorname*{arg\,max}_{q\in\mathcal{H},\boldsymbol{c}:\boldsymbol{c}=\boldsymbol{0}} H_d(q) \\ &= \exp\left(\sum_{k=1}^{d_\Theta} S_k\left(\boldsymbol{x},\boldsymbol{y}\right)\cdot\phi_k(\boldsymbol{\theta}^*) - \ln Z\left(\boldsymbol{\theta}^*|\boldsymbol{x}\right)\right) \end{aligned} \qquad [5.50]$$

with $Z\left(\boldsymbol{\theta}^*|\boldsymbol{x}\right)$ the normalization constant known as *partition function* and assuming that $h\left(\boldsymbol{x}\right)$, the base reference measure, to be a function only of $\boldsymbol{x}$. Moreover, $H_d\left(p_{\boldsymbol{\theta}}\right) \geq H_d\left(p_{\boldsymbol{\theta}^*}\right)$, $\forall \boldsymbol{\theta} \in \mathcal{H}_\Theta \subset \mathcal{H}$ and if $p_{\boldsymbol{\theta}} = p_{\boldsymbol{\theta}^*}$, then $H_d\left(p_{\boldsymbol{\theta}}\right) = H_d\left(p_{\boldsymbol{\theta}^*}\right)$

Theorem 5.4 states that the probability distribution that maximizes the entropy belongs to the exponential family. Moreover, the probability distribution that maximizes the entropy allows to describe the negative log-likelihood function as (Smola et al. 2005):

$$-\ln p_{\boldsymbol{\theta}^*}\left(\boldsymbol{y}|\boldsymbol{x}\right) = \ln Z\left(\boldsymbol{\theta}^*|\boldsymbol{x}\right) - \left\langle \boldsymbol{S}\left(\boldsymbol{x},\boldsymbol{y}\right), \boldsymbol{\phi}(\boldsymbol{\theta}^*)\right\rangle \qquad [5.51]$$

The conditional equipartition function $Z\left(\boldsymbol{\theta}|\boldsymbol{x}\right)$ reads:

$$Z\left(\boldsymbol{\theta}^*|\boldsymbol{x}\right) = \int_{\mathcal{Y}} \exp\left(\left\langle \boldsymbol{S}\left(\boldsymbol{x},\boldsymbol{y}\right), \boldsymbol{\phi}(\boldsymbol{\theta})\right\rangle\right) \mu(\boldsymbol{dy}) \qquad [5.52]$$

The gradient of the term $\ln Z\left(\boldsymbol{\theta}^*|\boldsymbol{x}\right)$ reads:

$$\boldsymbol{\nabla}_{\boldsymbol{\theta}} \ln Z\left(\boldsymbol{\theta}^*|\boldsymbol{x}\right) = \mathbb{E}_{(\boldsymbol{x},\boldsymbol{y})\sim p_{\boldsymbol{\theta}^*}}\left[\boldsymbol{S}\left(\boldsymbol{X},\boldsymbol{Y}\right)|\boldsymbol{X} = \boldsymbol{x}\right] \qquad [5.53]$$

which implies that conditional score reads[23]:

$$\begin{aligned} s_{\boldsymbol{Y}|\boldsymbol{X}=\boldsymbol{x}}\left(\boldsymbol{\theta}^*\right) = &\left(\boldsymbol{\phi}\left(\boldsymbol{\theta}^*\right) \otimes \boldsymbol{\nabla}_{\boldsymbol{\theta}}\right).\boldsymbol{S}\left(\boldsymbol{Y},\boldsymbol{x}\right) \\ &-\mathbb{E}_{(\boldsymbol{x},\boldsymbol{y})\sim p_{\boldsymbol{\theta}^*}}\left[\boldsymbol{S}\left(\boldsymbol{X},\boldsymbol{Y}\right)|\boldsymbol{X} = \boldsymbol{x}\right] \end{aligned} \qquad [5.54]$$

Moreover, the Fisher's information computed as the mean Hessian in equation [5.35] corresponds to the conditional covariance matrix of the $\boldsymbol{S}$ in the basis of the gradient $\left(\boldsymbol{\phi}\left(\boldsymbol{\theta}^*\right) \otimes \boldsymbol{\nabla}_{\boldsymbol{\theta}}\right)$ (Smola et al. 2005; Campagne 2022):

$$\begin{aligned} &\mathbb{I}^F_{\boldsymbol{Y}|\boldsymbol{X}=\boldsymbol{x}}\left(\boldsymbol{\theta}^*\right) = -\left(\boldsymbol{\phi}\left(\boldsymbol{\theta}^*\right) \otimes \boldsymbol{\nabla}_{\boldsymbol{\theta}}\right).\mathbb{C}_{(\boldsymbol{x},\boldsymbol{y})\sim p_{\boldsymbol{\theta}^*}} \\ &\times\left(\boldsymbol{S}\left(\boldsymbol{X},\boldsymbol{Y}\right) \otimes \boldsymbol{S}\left(\boldsymbol{X},\boldsymbol{Y}\right)|\boldsymbol{X}\right)\left(\boldsymbol{\nabla}_{\boldsymbol{\theta}} \otimes \boldsymbol{\phi}\left(\boldsymbol{\theta}^*\right)\right) \end{aligned} \qquad [5.55]$$

23 $\boldsymbol{\phi}\left(\boldsymbol{\theta}^*\right) \otimes \boldsymbol{\nabla}_{\boldsymbol{\theta}}$ is the gradient of $\boldsymbol{\phi}$.

REMARK 5.17.– The interesting aspect of equation [5.55] is that it depends only on the information provided by the sufficient statistics S_k.

The maximum entropy principle has a strong analogy with the Fisher information theory, since the value of the "best" estimator $\hat{\boldsymbol{\theta}}_N$ that maximizes the empirical conditional negative log-likelihood is asymptotically unbiased. Equation [5.54] adds the fact that $\hat{\boldsymbol{\theta}}_N$ makes the conditional statistics observations $\boldsymbol{S}(\boldsymbol{Y}, \boldsymbol{x})$ converge toward their average values that minimize the negative log-likelihood. Moreover, for a sufficiently large dataset $N \rightarrow +\infty$, the MLE $\hat{\boldsymbol{\theta}}_N$ tends to be normally distributed, that is, its probability distribution tends to belong to the exponential family. In other words, the MLE not only minimizes the negative log-likelihood but also maximizes the discrete conditional entropy of the underlying dataset $H(\boldsymbol{Y}|\boldsymbol{X}; \mathcal{D}_{\mathcal{XY}})$. In other words, the maximum entropy of a statistical model is reached by minimizing the negative log-likelihood of an exponential probability distribution (Campagne 2022). The quest for the best estimator can be seen as a minimax problem:

$$\hat{\boldsymbol{\theta}}_N = \underset{\boldsymbol{\theta} \in \Theta}{\arg\min} \max_{p_{\boldsymbol{\theta}} \in \mathcal{H}_\Theta} H\left(p_{\boldsymbol{\theta}}(Y|\boldsymbol{X})\right) \qquad [5.56]$$

5.3. Optimizing a neural network

The search for the "best" set of weights $\hat{\boldsymbol{\theta}}_N$ that minimizes the empirical loss function $\mathcal{L} : \mathbb{R}^{d_\Theta} \rightarrow \mathbb{R}$[24], according to equation [5.56], is an unconstrained optimization problem over finite-dimensional yet large Hilbert spaces $\mathbb{R}^{d_\Theta}$. This kind of problem is notoriously hard to solve, especially in non-convex setups (i.e. when neither the loss function nor the search space Θ are convex[25]). In practice, the risk of finding local non-unique minimizers $\hat{\boldsymbol{\theta}}_N$, for which $\mathcal{L}\left(\hat{\boldsymbol{\theta}}_N\right) \leq \mathcal{L}(\boldsymbol{\theta})$

24 The loss function $\mathcal{L}\left(\boldsymbol{h}_{\boldsymbol{\theta}}; \mathcal{D}_{\mathcal{XY}}\right)$ will be appearing in the remainder of the chapter as $\mathcal{L}\left(\boldsymbol{h}_{\boldsymbol{\theta}}\right)$, $\mathcal{L}(\boldsymbol{\theta})$.

25 Θ is convex if and only if for all $(\boldsymbol{\theta}_1, \boldsymbol{\theta}_2) \in \Theta^2$, $\forall t \in (0,1)$, $\boldsymbol{\theta}_t = t\boldsymbol{\theta}_1 + (1-t)\boldsymbol{\theta}_2 \in \Theta$.

only within an open neighborhood U of $\hat{\boldsymbol{\theta}}_N$, is quite recurrent in machine learning. In practice, the empirical loss function can be considered a *proper* function on $\mathbb{R}^{d_\Theta}$, that is, its non-empty domain of definition $\mathrm{dom}(\mathcal{L}) \subseteq \mathbb{R}^{d_\Theta}$. Moreover, $\mathcal{L}$ is generally *coercive*, $\lim_{\|\boldsymbol{\theta}\| \to +\infty} \mathcal{L}(\boldsymbol{\theta}) = +\infty$. In this case, one can define the search space Θ at the bounded lower level set that reads (Peyré 2020):

$$\Theta := \left\{\boldsymbol{\theta} \in \mathbb{R}^{d_\Theta} : \mathcal{L}(\boldsymbol{\theta}) \leq \mathcal{L}\left(\boldsymbol{\theta}^{(0)}\right)\right\} \qquad [5.57]$$

with $\boldsymbol{\theta}^{(0)}$ being the initial guess on the NN weights (see section 5.2.4.1 of volume 2 for weight initialization strategies). If Θ defined in equation [5.57] is closed, it is compact and the Bolzano–Weierstrass theorem assures that it exists at least one minimizer on it.

THEOREM 5.5 (**Bolzano-Weierstrass theorem (*Rein analytischer Beweis*, Bolzano, 1817 and Weierstrass, 1870)**).– Sufficient condition for the existence of a minimizer on a compact set

Given a function $\mathcal{L} : \Theta \to \mathbb{R}$ defined over a non-empty compact set $\Theta \subset \mathbb{R}^{d_\Theta}$, $\mathcal{L}$ proper and lower semi-continuous on Θ, then $\exists \hat{\boldsymbol{\theta}} \in \Theta$ such that:

$$\mathcal{L}\left(\hat{\boldsymbol{\theta}}\right) = \inf_{\boldsymbol{\theta} \in \Theta} \mathcal{L}(\boldsymbol{\theta}) = m \in \mathbb{R}$$

The Bolzano–Weierstrass theorem demands that $\mathcal{L}$ is lower semi-continuous (lsc), that is, if there exists a suite $(\boldsymbol{\theta}_n)_{n \in \mathbb{N}} \subset \Theta$ such that when $\lim_{n \to +\infty} \boldsymbol{\theta}_n = \boldsymbol{\theta}$, it holds that $\liminf_{n \to +\infty} \mathcal{L}(\boldsymbol{\theta}_n) \geq \mathcal{L}(\boldsymbol{\theta})$. Roughly speaking, lsc functions can be discontinuous with a jump at $\boldsymbol{\theta}_d$, but in a neighborhood of $\boldsymbol{\theta}_d$ the function is not much lower than $\mathcal{L}(\boldsymbol{\theta}_d)$. The epigraph of an lsc function is close. All continuous functions are lsc and so are the sum of lsc functions and $\sup_i \mathcal{L}_i$, with $\mathcal{L}_i$ lsc (Chouzenoux and Pesquet 2019). Continuous functions $\mathcal{L}$ imply that the lower level set Θ is closed (therefore compact): the Bolzano–Weierstrass applies. For domains Θ that are not necessarily closed, but for which $\mathcal{L}$ is coercive and lsc, one can prove that the set of minimizers $C = \arg\min_{\boldsymbol{\theta}\in\Theta}\mathcal{L}(\boldsymbol{\theta})$ is a non-empty compact set. If $\mathcal{L}$ is convex, then C is convex too, despite the possible existence of multiple minimizers.

The necessary condition for $\hat{\boldsymbol{\theta}}_N$ to be a local minimizer of $\mathcal{L}$ is provided by the first-order Peano–Kantorovitch condition, which reads:

THEOREM 5.6 (**Peano-Kantorovitch necessary condition for the existence of a local minimizer**).– If $\hat{\boldsymbol{\theta}}_N$ is a local minimizer of $\mathcal{L} : \Theta \to \mathbb{R}$ on $U \cap \Theta, U \in \mathcal{V}\left(\hat{\boldsymbol{\theta}}_N\right)$ and if $\mathcal{L}$ is differentiable in $\hat{\boldsymbol{\theta}}_N$, then:

$$\forall \delta\boldsymbol{\theta} \in T_{\hat{\boldsymbol{\theta}}_N}\Theta : d_{\hat{\boldsymbol{\theta}}_N}\mathcal{L}(\delta\boldsymbol{\theta}) \geq 0 \qquad [5.58]$$

or alternatively:

$$\boldsymbol{\nabla}_{\boldsymbol{\theta}}\mathcal{L}\left(\hat{\boldsymbol{\theta}}_N\right) \in \left(T_{\hat{\boldsymbol{\theta}}_N}\Theta\right)^+ \qquad [5.59]$$

with $d_{\hat{\boldsymbol{\theta}}_N}\mathcal{L}$ the differential of $T_{\hat{\boldsymbol{\theta}}_N}\Theta$ at $\hat{\boldsymbol{\theta}}_N$, representing the Bouligand's contingent cone (or set of tangent directions to Θ at $\hat{\boldsymbol{\theta}}_N$) defined as:

$$T_{\hat{\boldsymbol{\theta}}_N}\Theta := \overline{\bigcap_{\varepsilon>0}\left]\frac{1}{\varepsilon}, +\infty\right[\left(\Theta - \hat{\boldsymbol{\theta}}_N\right)}$$

and such that $\delta\boldsymbol{\theta} \in T_{\hat{\boldsymbol{\theta}}_N}\Theta \iff \exists (t_n)_{n\in\mathbb{N}} \subset \mathbb{R}, (\boldsymbol{\theta}_n)_{n\in\mathbb{N}} \subset \Theta$ such that:

$$\lim_{n\to+\infty} t_n \geq 0, \qquad \lim_{n\to+\infty}\frac{\boldsymbol{\theta}_n - \hat{\boldsymbol{\theta}}_N}{t_n} = \delta\boldsymbol{\theta}$$

$\left(T_{\hat{\boldsymbol{\theta}}_N}\Theta\right)^+$ is instead defined as the dual positive cone:

$$\left(T_{\hat{\boldsymbol{\theta}}_N}\Theta\right)^+ := \left\{\boldsymbol{\theta} \in \Theta : \langle\boldsymbol{\theta}, \delta\boldsymbol{\theta}\rangle \geq 0, \forall\delta\boldsymbol{\theta} \in T_{\hat{\boldsymbol{\theta}}_N}\Theta\right\}$$

If Θ is convex (i.e., $t\boldsymbol{\theta}_1 + (1-t)\boldsymbol{\theta}_2 \in \Theta$, $\forall (\boldsymbol{\theta}_1, \boldsymbol{\theta}_2) \in \Theta^2$ and $\forall t \in [0,1]$), the Peano–Kantorovitch inequality simply becomes:

$$\forall \boldsymbol{\theta} \in \Theta : d_{\hat{\boldsymbol{\theta}}_N} \mathcal{L}\left(\boldsymbol{\theta} - \hat{\boldsymbol{\theta}}_N\right) \geq 0 \qquad [5.60]$$

Equation [5.60] turns into a sufficient and necessary condition for $\hat{\boldsymbol{\theta}}_N$ to be a global minimum on Θ if $\mathcal{L}$ is convex. In practice, non-convex loss functions hinder the quest for the global minimum. Therefore, the following framework does not assume any a priori convexity of $\mathcal{L}$. However, the restriction to convex sets Θ practically holds and will adopted from now on. Moreover, in unconstrained problems, since the Bouligand's contingent cone of Θ reads $T_{\hat{\boldsymbol{\theta}}_N}\Theta = \Theta$ and the dual of Θ is $\{\mathbf{0}\}$, then the Peano–Kantorovitch inequality boils down to the Fermat's theorem:

THEOREM 5.7 (**Fermat necessary condition for the existence of a local minimizer**).– Given a function $\mathcal{L} : \Theta \to \mathbb{R}^{d_\Theta} \to \mathbb{R}$, $\mathcal{L}$ proper on Θ, if $\hat{\boldsymbol{\theta}}_N \in \Theta$ is a local minimum of $\mathcal{L}$ on the neighborhood V and $\mathcal{L}$ admits directional derivatives over $V \cap \Theta$, then:

$$D_{\boldsymbol{\theta}-\hat{\boldsymbol{\theta}}_N} \mathcal{L}\left(\hat{\boldsymbol{\theta}}_N; \mathcal{D}_{\mathcal{XY}}\right) = 0, \quad \forall \boldsymbol{\theta} \in \Theta \qquad [5.61]$$

where $D_{\boldsymbol{\theta}-\hat{\boldsymbol{\theta}}_N} \mathcal{L}\left(\hat{\boldsymbol{\theta}}_N; \mathcal{D}_{\mathcal{XY}}\right)$ represent the direction derivative of $\mathcal{L}$ in $\hat{\boldsymbol{\theta}}_N$ along the direction $\boldsymbol{\theta} - \hat{\boldsymbol{\theta}}_N$. Because of the Riesz–Fréchet theorem applied to the linear operator $D_{\boldsymbol{\theta}-\hat{\boldsymbol{\theta}}_N}\mathcal{L}$, this conditions boils down to:

$$\boldsymbol{\nabla}_{\boldsymbol{\theta}} \mathcal{L}\left(\hat{\boldsymbol{\theta}}_N; \mathcal{D}_{\mathcal{XY}}\right) = \mathbf{0} \qquad [5.62]$$

The zero gradient is not sufficient because the stationary point could be a minimum, maximum or saddle point. If $\mathcal{L}$ is convex, equation [5.62] is sufficient and necessary for a minimum to exist, but its uniqueness is not granted unless an even stronger condition is adopted for $\mathcal{L}$: to be strongly convex over a bounded convex set (Mula 2019).

REMARK 5.18.– $\mathcal{L}$ are generally the linear combinations of smooth (continuously differentiable) and non-smooth functions (e.g. the ℓ_1 regularization term). In non-smooth optimization, the gradient at certain points $\boldsymbol{\theta}$ is not well defined. In this case, the concept of *subderivative* and *subgradient* is introduced (Murphy 2022).

REMARK 5.19.– For a dataset $\mathcal{D}_{\mathcal{XY}}$ of N samples, the gradient of the empirical loss function reads:

$$\nabla_{\boldsymbol{\theta}}\mathcal{L}(\boldsymbol{\theta}) = \frac{1}{N}\sum_{(\boldsymbol{x}_k,\boldsymbol{y}_k)\in\mathcal{D}_{\mathcal{XY}}} \nabla_{\boldsymbol{\theta}}\ell\left(\boldsymbol{h}_\theta\left(\boldsymbol{x}_k\right), \boldsymbol{y}_k\right) + \nabla_{\boldsymbol{\theta}}\Omega\left(\boldsymbol{\theta}\right) \qquad [5.63]$$

The cost of computing the exact gradient in equation [5.63], by averaging across the whole dataset instances and then updating the weights (the so-called Batch Gradient Descent), is $\mathcal{O}(N \cdot d_\Theta)$. This computational cost is affordable for rather small datasets and shallow *NN*, being tailored only for convex and rather smooth loss functions. However, for very large datasets, the floating-point error induced by computing a sum over a large number of addends introduces spurious errors that can make the gradient descent algorithm fail.

The optimization of a *NN* is performed, in `PyTorch`, with the help of the library `torch.optim`[26]. The whole optimization algorithm takes only a few lines of code:

```
import torch

# Load the dataset
DXY = ...

# Define the NN
h_theta = torch.nn(...)

# Defin the learning rate
eta_max = 0.01
eta_min = 0.001
```

26 Available at: https://pytorch.org/docs/stable/optim.html.

```
# Define the number of epochs
n_e = 1200

# Choose the optimizer (SGD in this case)
optimizer = torch.optim.SGD(h_theta.parameters(), lr=eta)

# Define a triangular cyclic learning rate scheduling
scheduler = torch.optim.lr_scheduler.CyclicLR(optimizer,
        eta_min, eta_max, step_size_up=100,
          mode='triangular')

# Define loss function (MLE for regression)
loss_fn = torch.nn.MSELoss()

# iterate over the training epochs
for i in range(n_e)
        for xk, yk in DXY:
            def closure():
                optimizer.zero_grad()
                h_theta_xk = h_theta(xk)
                L = loss_fn(h_theta_xk, yk)
                L.backward()
                return L
            optimizer.step(closure)
```

A PyTorch optimizer is fed with an *iterable* containing the model parameters `h_theta.parameters()` to optimize based on the definition of a loss function (in the example above, the loss function corresponds to the MSE, provided by the `torch.nn` library, `loss_fn = torch.nn.MSELoss()`. The optimizer will hold the memory address of each parameter in the list `h_theta.parameters()`, having access each time to their current values and updating them (`optimizer.step(closure)`) based on the computed gradients (computed by the line `L.backward()`). Optimizer-specific options such as the learning rate `eta` are fed as additional arguments, such as a learning rate scheduling (in the example above, a triangular cyclic scheduling, bounded by `eta_min` and `eta_max`, with `step_size_up=100` training iterations in the increasing half of a cycle). PyTorch offers the possibility of parameter-wise options, which is activated by passing to the optimizer an iterable of dictionaries:

```
optim.SGD([{'params': h_theta.base.parameters()},
           {'params': h_theta.classifier.parameters(),
                 'lr': eta_max}
           ], lr=eta_min)
```

This means that `h_theta.base` parameters will use the default learning rate `eta_min`, whereas `h_theta.classifier` will be updated using a learning rate of `eta_max`.

The line `optimizer.step()` updates the gradients according to the SGD *delta rule* (see section 5.3.4). The line `optimizer.zero_grad()` resets the gradients previously computed and stored at each parameter as `.grad()` attribute. This attribute is `None` by default until it becomes a `Tensor` (the fundamental brick of the `PyTorch` data structure, storing multi-dimensional arrays and several attributes such as `type` and `device`, to assess whether it is stored on CPU or GPU), the first time a call to `backward()` computes gradients. The attribute will then contain the gradients computed and future calls to `backward()` will accumulate (add) gradients into it.

REMARK 5.20.– The standard optimization practice employed in deep learning makes use of single-precision floating-point format (i.e. numbers are encoded in a 32-bit-long – 4 bytes – binary code) and even half-precision floating-point format (i.e. numbers are encoded in a 16-bit-long (2 bytes) binary code). This choice is justified by the fact that gradient descent algorithms are run on GPU, primarily designed for interactive rendering, not computation. Compared to CPU, GPU have limited cache memory (where to store data to be repeatedly accessed) but higher sequential memory (where to store data that are naturally structured in a contiguous way, such as in an array). Therefore, a GPU is extremely fast and efficient in computing linear algebra operations on contiguously stored data, which the CPU feeds it continuously with. However, to reach peak performances, those linear algebra operations (such as the gradient descent one) are performed in single- or even half-precision floating-point format. The double-precision floating-point format can hinder GPU due to its limited memory capacity, which is why it is rarely used. However, this choice implies possible divergence of the Batch Gradient Descent algorithm on large datasets.

5.3.1. *First-order gradient descent for empirical loss minimization*

The quest for $\hat{\boldsymbol{\theta}}_N$ is iteratively conducted by adopting the equation [5.60], that is, searching for a local minimizer of the empirical loss function $\mathcal{L}$ along the steepest descent direction $-\boldsymbol{\nabla}_{\boldsymbol{\theta}}\mathcal{L}$. In this sense, adopting the Peano–Kantorovitch inequality, it holds that:

$$\boldsymbol{\theta}^{(i)} = \hat{\boldsymbol{\theta}}_N + \eta \boldsymbol{\nabla}_{\boldsymbol{\theta}}\mathcal{L}\left(\hat{\boldsymbol{\theta}}_N\right) \qquad [5.64]$$

with $\eta \in \mathbb{R}^+$ being a sufficiently small *learning rate*, for which

$$\mathcal{L}\left(\boldsymbol{\theta}^{(i)} - \eta \cdot \boldsymbol{\nabla}_{\boldsymbol{\theta}}\mathcal{L}\left(\boldsymbol{\theta}^{(i)}\right)\right) \geq \mathcal{L}\left(\hat{\boldsymbol{\theta}}_N\right)$$

Equation [5.64] suggests an iterative strategy, called *delta rule*, to update the weights from iteration i to iteration $i+1$:

$$\boldsymbol{\theta}^{(i+1)} = \boldsymbol{\theta}^{(i)} - \eta^{(i)} \boldsymbol{\nabla}_{\boldsymbol{\theta}}\mathcal{L}\left(\boldsymbol{\theta}^{(i)}\right), \quad \eta^{(i)} \in \mathbb{R}^+ \qquad [5.65]$$

When the norm of the gradient $\|\boldsymbol{\nabla}_{\boldsymbol{\theta}}\mathcal{L}\left(\boldsymbol{\theta}^{(i)}\right)\|$ is sufficiently small, the iterative strategy stops because a (local) minimizer has been reached. The *delta rule* is directly issued from equation [5.60] and from the Cauchy–Schwartz inequality, the latter implying that within a neighborhood $U \in \mathcal{V}\left(\hat{\boldsymbol{\theta}}_N\right)$ of $\hat{\boldsymbol{\theta}}_N$, it holds that:

$$\|\boldsymbol{\theta}^{(i)} - \hat{\boldsymbol{\theta}}_N\| \geq \left\langle \frac{\boldsymbol{\nabla}_{\boldsymbol{\theta}}\mathcal{L}\left(\hat{\boldsymbol{\theta}}_N\right)}{\|\boldsymbol{\nabla}_{\boldsymbol{\theta}}\mathcal{L}\left(\hat{\boldsymbol{\theta}}_N\right)\|}, \boldsymbol{\theta}^{(i)} - \hat{\boldsymbol{\theta}}_N \right\rangle \geq 0,$$

$$\forall \boldsymbol{\theta}^{(i)} \in U \cap \Theta \qquad [5.66]$$

In this sense, the negative gradient represents a greedy steepest descent direction, because to minimize the function $\mathcal{L}\left(\boldsymbol{\theta}^{(i)} + \boldsymbol{\delta\theta}^{(i)}\right)$, with the smallest increment $\boldsymbol{\delta\theta}^{(i)}$, one should minimize the function

$g\left(\boldsymbol{\delta\theta}^{(i)}\right) = \mathcal{L}\left(\boldsymbol{\theta}^{(i)} + \boldsymbol{\delta\theta}^{(i)}\right) + \frac{c}{2}\|\boldsymbol{\delta\theta}^{(i)}\|^2$, which corresponds to the solution of the equation $\boldsymbol{\nabla}_{\boldsymbol{\theta}}\mathcal{L}\left(\boldsymbol{\theta}^{(i)} + \boldsymbol{\delta\theta}^{(i)}\right) = -c\boldsymbol{\delta\theta}^{(i)}$. In order for the update to seek for a lower loss value:

$$\mathcal{L}\left(\boldsymbol{\theta}^{(i)} + \boldsymbol{\delta\theta}^{(i)}\right) = \mathcal{L}\left(\boldsymbol{\theta}^{(i)}\right) + \left\langle \boldsymbol{\nabla}_{\boldsymbol{\theta}}\mathcal{L}\left(\boldsymbol{\theta}^{(i)}\right), \boldsymbol{\delta\theta}^{(i)}\right\rangle + o\left(\|\boldsymbol{\delta\theta}^{(i)}\|\right) \leq \mathcal{L}\left(\boldsymbol{\theta}^{(i)}\right),$$

the iterative increment $\boldsymbol{\delta\theta}^{(i)}$ must point toward the direction opposite to the gradient (Murphy 2022):

$$\lim_{\|\boldsymbol{\delta\theta}^{(i)}\|\to 0} \left\langle \frac{\boldsymbol{\delta\theta}^{(i)}}{\|\boldsymbol{\delta\theta}^{(i)}\|}, \boldsymbol{\nabla}_{\boldsymbol{\theta}}\mathcal{L}\left(\boldsymbol{\theta}^{(i)}\right)\right\rangle + \frac{o\left(\|\boldsymbol{\delta\theta}^{(i)}\|\right)}{\|\boldsymbol{\delta\theta}^{(i)}\|} \leq 0$$

The previous limit implies that $c > 0$ and since the gradient is continuous by assumption, it holds that (Peyré 2020):

$$\arg\min_{\|\boldsymbol{\delta\theta}^{(i)}\|=1} \mathcal{L}\left(\boldsymbol{\theta}^{(i)} + \boldsymbol{\delta\theta}^{(i)}\right) = -\frac{\boldsymbol{\nabla}_{\boldsymbol{\theta}}\mathcal{L}\left(\boldsymbol{\theta}^{(i)}\right)}{\|\boldsymbol{\nabla}_{\boldsymbol{\theta}}\mathcal{L}\left(\boldsymbol{\theta}^{(i)}\right)\|} \qquad [5.67]$$

In other words, the greediest update direction is provided by the direction in the unit ball that extends the farthest in the direction $-\boldsymbol{\nabla}_{\boldsymbol{\theta}}\mathcal{L}\left(\boldsymbol{\theta}^{(i)}\right)$ (Boyd and Vandenberghe 2023).

5.3.1.1. *Learning rate scheduling*

The choice of $\eta^{(i)}$ is delicate. The learning rate can be efficiently updated at each iteration, for instance, according to the greedy choice provided by the classical (exact) *line-search* method:

$$\bar{\eta}^{(i)} = \arg\min_{\eta>0} \mathcal{L}\left(\boldsymbol{\theta}^{(i)} - \eta \cdot \boldsymbol{\delta\theta}^{(i)}\right) \qquad [5.68]$$

Exact line-search methods (obtained by first-order Taylor's expansion of $f(\eta) = \mathcal{L}\left(\boldsymbol{\theta}^{(i)} - \eta \cdot \boldsymbol{\delta\theta}^{(i)}\right)$) yields weight updates orthogonal to the steepest descent, as expressed:

$$\left\langle \boldsymbol{\nabla}_{\boldsymbol{\theta}}\mathcal{L}\left(\boldsymbol{\theta}^{(i)} - \bar{\eta}^{(i)}\boldsymbol{\delta\theta}^{(i)}\right), \boldsymbol{\delta\theta}^{(i)}\right\rangle = 0 \quad [5.69]$$

The exact line-search greedy choice is rather demanding from a computational standpoint since it is the solution of the minimization problem in equation [5.68]. However, there are cheaper alternatives, such as the *Armijo backtracking method* (inexact line-search), which approximates the solution of equation [5.68] by iteratively decreasing the initial learning rate $\eta^{(0)}$, or the maximum one $\eta_{\max} = \max_{0 \le i < n_e} \eta^{(i)}$, named $\bar{\eta}$ in the following, by factor $\xi \in \left]0, 1\right[$ until the *Armijo–Goldstein test* is passed, that is, until iteration k, for which the differential of the loss gets smaller than a certain tolerance ϵ, according to the following condition (Murphy 2022):

$$\mathcal{L}\left(\boldsymbol{\theta}^{(i)} + \xi^k\bar{\eta}\boldsymbol{\delta\theta}^{(i)}\right) \le \mathcal{L}\left(\boldsymbol{\theta}^{(i)}\right) - \epsilon\xi^k\bar{\eta}\left\langle \boldsymbol{\nabla}_{\boldsymbol{\theta}}\mathcal{L}\left(\boldsymbol{\theta}^{(i)}\right), \boldsymbol{\delta\theta}^{(i)}\right\rangle \quad [5.70]$$

with $\epsilon = 10^{-4}$. In general, the learning rate update is scheduled, usually with a decay formulated as follows (Peyré 2020):

$$\eta^{(i)} = \frac{\eta_{\max}}{1 + i \cdot \eta_d} \quad [5.71]$$

with η_d being called the *learning rate decay*. However, it is suggested to start with a small learning rate $\eta^{(0)}$, linearly increasing in a "warm-up" phase, until it reaches its maximum at $i = i_0$, as shown in Figure 5.6, and then decreased according to equation [5.71] ($\eta^{(i)} = \frac{\eta^{(i_0)}}{1+(i-i_0)\eta_d}$). The warm-up phase is necessary for practical reasons: the initial iterations yield untrusted gradient values $\boldsymbol{\nabla}_{\boldsymbol{\theta}}\mathcal{L}$, which strongly depend on the initialization of the weights at $i = 0$ (see section 5.2.4). Adaptive optimizers (e.g. Adam and RMSProp, see section 5.3.7.4 and section 5.3.7.2) are especially sensitive to those initial untrusted estimates of the gradient statistics.

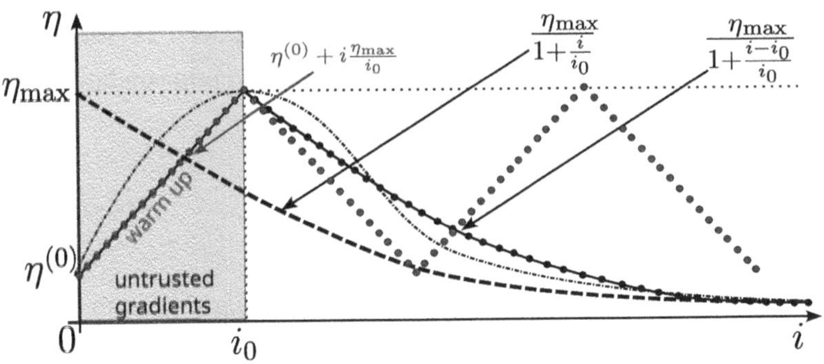

Figure 5.6. *Different learning rate scheduling strategies. Different decay scheduling rules are shown, with an initial linear warm-up phase. Red markers indicate the Cyclic Learning Rate Scheduler proposed by Smith (2017). Modified after the* Deep Learning Optimisée sur Jean Zay (DLO-JZ): documentation *(see http://www.idris.fr/formations/dlo-jz/)*

Another scheduling strategy, called Cyclic Learning Rate Scheduler, was proposed by Smith (2017) (see red line in Figure 5.6). This choice implies to first calibrate the lower and upper bounds $\eta_{\min}$ and $\eta_{\max}$, respectively (if not known beforehand), and then to reiterate over several linear loops. However, Smith (2017) seemingly suggests that one cycle could be enough. Other examples of learning rate schedulers are piecewise constant, exponential and polynomial decay, cosine annealing and even learning rate schedulers that adaptively reduce the learning rate when a metric has stopped improving, to improve the convergence by reducing η by a factor of 2-10 if after a certain number of epochs (*patience*) the metric does not change significantly. Theoretically, the convergence of first-order gradient-descent methods with a learning rate scheduler is achieved if and only if the Robbins–Monro conditions are satisfied (Murphy 2022), that is, if $\lim\limits_{i\to+\infty} \eta^{(i)} = 0$ and $\lim\limits_{i\to+\infty} \frac{\sum_{i=0}^{+\infty}\left(\eta^{(i)}\right)^2}{\sum_{i=0}^{+\infty}\eta^{(i)}} = 0$.

5.3.1.2. *On the convergence of first-order gradient-descent methods*

If $\mathcal{L}$ is convex, a local minimizer exists. In this case, the learning rate can be bounded, provided that the gradient $\nabla_\theta \mathcal{L}$ is β-Lipschitz, that is, that $\|\nabla_{\boldsymbol{\theta}}\mathcal{L}(\boldsymbol{\theta}+\delta\boldsymbol{\theta}) - \nabla_{\boldsymbol{\theta}}\mathcal{L}(\boldsymbol{\theta})\| \leq \beta\|\delta\boldsymbol{\theta}\|$. Moreover, under such assumptions (convexity and β-Lipschitz regularity of the gradient), the first-order gradient descent converges, as stated by the following theorem by Nesterov (1983).

THEOREM 5.8 (**Convergence analysis of functions with Lipschitz gradients**; Nesterov 1983; Peyré 2020).– A convex function $\mathcal{L} : \Theta \to \mathbb{R}$, defined over a non-empty convex set $\Theta \subset \mathbb{R}^{d_\Theta}$, $\mathcal{L}$ proper on Θ, with $\mathcal{L} \in \mathcal{C}^1(\Theta)$ and a gradient $\nabla_{\boldsymbol{\theta}}\mathcal{L}$ is β-Lipschitz on (Θ). If $\exists \eta_{\min}, \eta_{\max} \in \mathbb{R}_+^*$ such that:

$$0 < \frac{1}{\beta} = \eta_{\min} \leq \eta^{(i)} < \eta_{\max} = \frac{2}{\beta}$$

then $\exists (\boldsymbol{\theta}^{(i)})_{i\in\mathbb{N}} \in \Theta$ such that $\boldsymbol{\theta}^{(i+1)} = \boldsymbol{\theta}^{(i)} - \eta^{(i)}\nabla_{\boldsymbol{\theta}}\mathcal{L}\left(\boldsymbol{\theta}^{(i)}\right)$ that converges to $\hat{\boldsymbol{\theta}}_N = \underset{\boldsymbol{x}\in\Theta}{\arg\min}\, \mathcal{L}(\boldsymbol{\theta}; \mathcal{D}_{\mathcal{XY}})$ and there exists a constant $A > 0$ such that:

$$\mathcal{L}\left(\boldsymbol{\theta}^{(n_e)}\right) - \mathcal{L}\left(\hat{\boldsymbol{\theta}}_N\right) \leq \frac{A}{n_e + 1} \qquad [5.72]$$

with n_e number of iterations of the *delta rule* (number of *epochs*).

A simple proof of the theorem is provided by (Peyré 2020) and it is based on the $\mathcal{L}$ Taylor's expansion with integral reminder, which reads:

$$\mathcal{L}(\boldsymbol{\theta}+\delta\boldsymbol{\theta}) - \mathcal{L}(\boldsymbol{\theta}) = \left\langle \nabla_{\boldsymbol{\theta}}\mathcal{L}(\boldsymbol{\theta}), \delta\boldsymbol{\theta} \right\rangle + \int_0^1 \left\langle \nabla_{\boldsymbol{\theta}}\mathcal{L}(\boldsymbol{\theta}_t) - \nabla_{\boldsymbol{\theta}}\mathcal{L}(\boldsymbol{\theta}), \delta\boldsymbol{\theta} \right\rangle dt$$

with $\boldsymbol{\theta}_t = \boldsymbol{\theta} + t\delta\boldsymbol{\theta} \in \Theta$ (convex). Based on Taylor's expansion and given the β-Lipschitz regularity of its gradient, $\mathcal{L}$ can be bounded as follows:

$$\mathcal{L}\left(\boldsymbol{\theta}^{(i+1)}\right) \leq \mathcal{L}\left(\boldsymbol{\theta}^{(i)}\right) + \left\langle \nabla_{\boldsymbol{\theta}}\mathcal{L}\left(\boldsymbol{\theta}^{(i)}\right), \delta\boldsymbol{\theta}^{(i)} \right\rangle + \frac{\beta}{2}\|\delta\boldsymbol{\theta}^{(i)}\|^2 \qquad [5.73]$$

Thanks to the continuity of $\mathcal{L}$, the Cauchy–Schwartz inequality implies that:

$$\mathcal{L}(\boldsymbol{\theta}+\boldsymbol{\delta\theta})-\mathcal{L}(\boldsymbol{\theta})\leq\left\langle\boldsymbol{\nabla}_{\boldsymbol{\theta}}\mathcal{L}(\boldsymbol{\theta}),\boldsymbol{\delta\theta}\right\rangle+\frac{\beta}{2}\|\boldsymbol{\delta\theta}\|^2 \qquad [5.74]$$

Replacing the *delta rule* (equation [5.65]) in equation [5.73], we obtain the following expression:

$$\mathcal{L}\left(\boldsymbol{\theta}^{(i+1)}\right)\leq\mathcal{L}\left(\boldsymbol{\theta}^{(i)}\right)+\eta^{(i)}\left(\frac{\eta^{(i)}\cdot\beta}{2}-1\right)\|\boldsymbol{\nabla}_{\boldsymbol{\theta}}\mathcal{L}\left(\boldsymbol{\theta}^{(i)}\right)\|^2 \qquad [5.75]$$

which requires an upper bound for the learning rate, that is, $\eta^{(i)}<\frac{2}{\beta}=\eta_{\max}$, so to obtain $\mathcal{L}\left(\boldsymbol{\theta}^{(i+1)}\right)\leq\mathcal{L}\left(\boldsymbol{\theta}^{(i)}\right)$ and therefore $\tau^{(i)}=\eta^{(i)}\left(\frac{\eta^{(i)}\beta}{2}-1\right)<0$. To find the lower bound $\eta_{\min}$, one must assume convexity of the loss function $\mathcal{L}$, for which it holds that[27]:

$$\mathcal{L}\left(\hat{\boldsymbol{\theta}}_N\right)\geq\mathcal{L}\left(\boldsymbol{\theta}^{(i)}\right)+\left\langle\boldsymbol{\nabla}_{\boldsymbol{\theta}}\mathcal{L}\left(\boldsymbol{\theta}^{(i)}\right),\hat{\boldsymbol{\theta}}_N-\boldsymbol{\theta}^{(i)}\right\rangle \qquad [5.76]$$

Comparing equation [5.75] and equation [5.76], replacing the *delta rule* into it and considering that $\tau^{(i)}<0$ leads to:

$$\begin{aligned}\mathcal{L}\left(\boldsymbol{\theta}^{(i+1)}\right)&\leq\mathcal{L}\left(\hat{\boldsymbol{\theta}}_N\right)+\left\langle\boldsymbol{\nabla}_{\boldsymbol{\theta}}\mathcal{L}\left(\boldsymbol{\theta}^{(i)}\right),\boldsymbol{\theta}^{(i)}-\hat{\boldsymbol{\theta}}_N\right\rangle\\&+\tau^{(i)}\|\boldsymbol{\nabla}_{\boldsymbol{\theta}}\mathcal{L}\left(\boldsymbol{\theta}^{(i)}\right)\|^2\\&=\mathcal{L}\left(\hat{\boldsymbol{\theta}}_N\right)+\frac{\|\boldsymbol{\theta}^{(i)}-\hat{\boldsymbol{\theta}}_N+2\tau^{(i)}\boldsymbol{\nabla}_{\boldsymbol{\theta}}\mathcal{L}\left(\boldsymbol{\theta}^{(i)}\right)\|^2-\|\boldsymbol{\theta}^{(i)}-\hat{\boldsymbol{\theta}}_N\|^2}{4\tau^{(i)}}\end{aligned} \qquad [5.77]$$

27 Convexity of a function $\mathcal{L}$ is provided by the following property: $\mathcal{L}(\boldsymbol{\theta}+t\boldsymbol{\delta\theta})\leq(1-t)\mathcal{L}(\boldsymbol{\theta})+t\mathcal{L}(\boldsymbol{\theta}+\boldsymbol{\delta\theta})$. If $\mathcal{L}$ is differentiable and C^1, then the Taylor's expansion at first-order reads $\mathcal{L}(\boldsymbol{\theta}+t\boldsymbol{\delta\theta})=\mathcal{L}(\boldsymbol{\theta})+t\left\langle\boldsymbol{\nabla}_{\boldsymbol{\theta}}\mathcal{L},\boldsymbol{\delta\theta}\right\rangle+o(t)\leq(1-t)\mathcal{L}(\boldsymbol{\theta})+t\mathcal{L}(\boldsymbol{\theta}+\boldsymbol{\delta\theta})$, which implies that:

$$\mathcal{L}(\boldsymbol{\theta})+\left\langle\boldsymbol{\nabla}_{\boldsymbol{\theta}}\mathcal{L},\boldsymbol{\delta\theta}\right\rangle+\lim_{t\to0}\frac{o(t)}{t}\leq\mathcal{L}(\boldsymbol{\theta}+\boldsymbol{\delta\theta})$$

$$= \mathcal{L}\left(\hat{\boldsymbol{\theta}}_N\right) + \frac{\|\boldsymbol{\theta}^{(i)} - \hat{\boldsymbol{\theta}}_N\|^2 - \|\boldsymbol{\theta}^{(i)} - \hat{\boldsymbol{\theta}}_N + 2\tau^{(i)}\boldsymbol{\nabla}_{\boldsymbol{\theta}}\mathcal{L}\left(\boldsymbol{\theta}^{(i)}\right)\|^2}{4|\tau^{(i)}|}$$

$$= \mathcal{L}\left(\hat{\boldsymbol{\theta}}_N\right) \qquad [5.78]$$

$$+ \frac{\|\boldsymbol{\theta}^{(i)} - \hat{\boldsymbol{\theta}}_N\|^2 - \|\boldsymbol{\theta}^{(i+1)} - \hat{\boldsymbol{\theta}}_N + \eta^{(i)}\left(\eta^{(i)}\beta - 1\right)\boldsymbol{\nabla}_{\boldsymbol{\theta}}\mathcal{L}\left(\boldsymbol{\theta}^{(i)}\right)\|^2}{4|\tau^{(i)}|}$$

From equation [5.78], it must be noted that, to assure that $\mathcal{L}\left(\boldsymbol{\theta}^{(i+1)}\right) \geq \mathcal{L}\left(\hat{\boldsymbol{\theta}}_N\right)$, one must assure that:

$$\begin{aligned} \|\boldsymbol{\theta}^{(i)} - \hat{\boldsymbol{\theta}}_N\| &\geq \|\boldsymbol{\theta}^{(i+1)} - \hat{\boldsymbol{\theta}}_N + \eta^{(i)}\left(\eta^{(i)}\beta - 1\right)\boldsymbol{\nabla}_{\boldsymbol{\theta}}\mathcal{L}\left(\boldsymbol{\theta}^{(i)}\right)\| \\ &\geq \left|\|\boldsymbol{\theta}^{(i+1)} - \hat{\boldsymbol{\theta}}_N\| - \eta^{(i)}|\eta^{(i)}\beta - 1|\|\boldsymbol{\nabla}_{\boldsymbol{\theta}}\mathcal{L}\|\right| \end{aligned} \qquad [5.79]$$

and due to the Lipschitz property:

$$\left(\eta^{(i)}\beta|\eta^{(i)}\beta - 1| + 1\right) \cdot \|\boldsymbol{\theta}^{(i)} - \hat{\boldsymbol{\theta}}_N\| \geq \|\boldsymbol{\theta}^{(i+1)} - \hat{\boldsymbol{\theta}}_N\| \qquad [5.80]$$

which implies that $\eta_{\min} = \frac{1}{\beta}$, to achieve convergence (Peyré 2020). Moreover, summing equation [5.78] over the total number of epochs n_e, we obtain:

$$n_e\left(\mathcal{L}\left(\boldsymbol{\theta}^{(n_e-1)}\right) - \mathcal{L}\left(\hat{\boldsymbol{\theta}}_N\right)\right) \leq \sum_{i=0}^{n_e-1}\mathcal{L}\left(\boldsymbol{\theta}^{(i+1)}\right) - n_e \cdot \mathcal{L}\left(\hat{\boldsymbol{\theta}}_N\right)$$

$$[5.81]$$

$$\leq \sum_{i=0}^{n_e}\frac{\|\boldsymbol{\theta}^{(i)} - \hat{\boldsymbol{\theta}}_N\|^2 - \|\boldsymbol{\theta}^{(i+1)} - \hat{\boldsymbol{\theta}}_N + \eta^{(i)}\left(\eta^{(i)}\beta - 1\right)\boldsymbol{\nabla}_{\boldsymbol{\theta}}\mathcal{L}\left(\boldsymbol{\theta}^{(i)}\right)\|^2}{4|\tau_k|}$$

If, for instance, one selects the lowest learning rate possible, $\eta = \eta_{\min} = \frac{1}{\beta}$, then:

$$\mathcal{L}\left(\boldsymbol{\theta}^{(n_e-1)}\right) - \mathcal{L}\left(\hat{\boldsymbol{\theta}}_N\right) \leq \frac{\beta}{2n_e} \sum_{i=0}^{n_e-1} ||\boldsymbol{\theta}^{(i)} - \hat{\boldsymbol{\theta}}_N||^2 - ||\boldsymbol{\theta}^{(i+1)} - \hat{\boldsymbol{\theta}}_N||^2 \qquad [5.82]$$

$$= \frac{2}{n_e\beta}\left(||\boldsymbol{\theta}_0 - \hat{\boldsymbol{\theta}}_N||^2 - ||\boldsymbol{\theta}_{n_e} - \hat{\boldsymbol{\theta}}_N||^2\right) \leq \frac{\beta||\boldsymbol{\theta}_0 - \hat{\boldsymbol{\theta}}_N||^2}{2n_e}$$

Equation [5.82] proves equation [5.72] (Peyré 2020).

5.3.2. *Second-order gradient descent methods*

If we consider a quadratic loss function $\mathcal{L} = \frac{1}{2}||\mathbb{A}\boldsymbol{\theta} - \boldsymbol{b}||^2$, its single global minimizer $\hat{\boldsymbol{\theta}}$ is equal to $\hat{\boldsymbol{\theta}} = \left(\mathbb{A}^T\mathbb{A}\right)^{-1}\mathbb{A}^T\boldsymbol{b}$, with $\left(\mathbb{A}^T\mathbb{A}\right)^{-1}\mathbb{A}^T$ the Moore–Penrose pseudo-inverse of $\mathbb{A}$ (which is not necessarily square nor symmetric, see the *normal equation* in equation [2.9]). $\hat{\boldsymbol{\theta}}$ is a unique solution of the system $\mathbb{A}^T\mathbb{A}\boldsymbol{\theta} = \mathbb{A}^T\boldsymbol{b}$, provided that the linear application $\mathbb{A}^T\mathbb{A}$ is bijective and therefore that $\mathbb{A}^T\mathbb{A}\boldsymbol{\theta} = \mathbf{0}$ if and only if $\boldsymbol{\theta} = \mathbf{0}$. In such a case, the gradient of the loss function reads $\boldsymbol{\nabla}_{\boldsymbol{\theta}}\mathcal{L} = \mathbb{A}^T(\mathbb{A}\boldsymbol{\theta} - \boldsymbol{b})$. Moreover, the Moore–Penrose pseudo-inverse exists and $\left\langle\mathbb{A}^T\mathbb{A}\boldsymbol{\theta}, \boldsymbol{\theta}\right\rangle = ||\mathbb{A}\boldsymbol{\theta}||^2 = 0$ implies that $\mathbb{A}\boldsymbol{\theta} = \mathbf{0}$. Therefore, the global unique minimizer $\hat{\boldsymbol{\theta}}$ of a quadratic loss function exists if and only if $\ker(\mathbb{A}) = \{\mathbf{0}\}$ (see section 2.2 for the implications in linear regression). $\mathbb{A}^T\mathbb{A}$ corresponds to the Hessian of the loss function $\mathbb{H}_{\mathcal{L}}$, which is a linear symmetric operator ($\mathbb{H}_{\mathcal{L}} \in \mathcal{S}_{d_\Theta}(\mathbb{R})$), whose dual norm is defined as $||\mathbb{H}_{\mathcal{L}}||_* = \sup\limits_{\boldsymbol{\theta}\neq\mathbf{0}} \frac{||\mathbb{H}_{\mathcal{L}}\boldsymbol{\theta}||}{\boldsymbol{\theta}}$. Calling $\lambda_i(\mathbb{H})$ its eigenvalues, $\mathbb{H}_{\mathcal{L}}$ diagonalizes into the orthogonal basis of its eigenvectors ϕ_k, such that $\mathbb{H}_{\mathcal{L}} = \sum\limits_{k=1}^{d_\Theta} \lambda_k\phi_k \otimes \phi_k$. Therefore, $||\mathbb{H}_{\mathcal{L}}||_* = |\lambda_{\max}(\mathbb{H}_{\mathcal{L}})|$. The Hessian is positive definite if all its eigenvalues are strictly positive. The latter condition corresponds to $0 < \langle\mathbb{H}_{\mathcal{L}}\boldsymbol{\delta\theta}, \boldsymbol{\delta\theta}\rangle \leq ||\mathbb{H}_{\mathcal{L}}\boldsymbol{\delta\theta}|| \cdot ||\boldsymbol{\delta\theta}||$, for all $\boldsymbol{\delta\theta} \in \Theta \setminus \{\mathbf{0}\}$, synthetically written as $\mathbb{H}_{\mathcal{L}} \succ 0$.

For quadratic functions $\mathcal{L} = \frac{1}{2}||\mathbb{A}\boldsymbol{\theta} - \boldsymbol{b}||^2$, with $\ker(\mathbb{A}) = \{\mathbf{0}\}$, the first-order gradient descent method converges linearly, provided that

$0 < \alpha \leq \lambda_{\min}(\mathbb{H}_{\mathcal{L}}) \leq \lambda_{\max}(\mathbb{H}_{\mathcal{L}}) \leq \beta$, according to the following inequality (see Bertsekas 2015; Murphy 2022, for the proof):

$$\|\boldsymbol{\theta}^{(i+1)} - \hat{\boldsymbol{\theta}}_N\| \leq \left(\frac{\frac{\beta}{\alpha} - 1}{\frac{\beta}{\alpha} + 1}\right)^2 \cdot \|\left(\boldsymbol{\theta}^{(0)}\right) - \left(\hat{\boldsymbol{\theta}}_N\right)\| \qquad [5.83]$$

Such a condition is realized by adopting a *delta rule* with constant learning rate $\eta = \frac{2}{\beta+\alpha}$. The upper bound on the Hessian's eigenvalues, according to equation [5.74], corresponds to a quadratic loss function with β-Lipschitz gradient:

$$\|\boldsymbol{\nabla}_{\boldsymbol{\theta}}\mathcal{L}(\boldsymbol{\theta} + \boldsymbol{\delta\theta}) - \boldsymbol{\nabla}_{\boldsymbol{\theta}}\mathcal{L}(\boldsymbol{\theta})\| = \|(\mathbb{A}_T\mathbb{A})\,\boldsymbol{\delta\theta}\| = \|\mathbb{H}_{\mathcal{L}}\boldsymbol{\delta\theta}\| \leq \beta\|\boldsymbol{\delta\theta}\| \qquad [5.84]$$

which implies a bounded (therefore continuous) Hessian linear operator $\|\mathbb{H}_{\mathcal{L}}\|_* \leq \beta$. This result is in agreement with the choice of a learning rate $\eta = \frac{2}{\beta+\alpha} \in \left[\frac{1}{\beta}, \frac{2}{\beta}\right[$, as proved in theorem 5.8, that in this case apply to a quadratic function with eigenvalues bounded by above by $\beta > 0$.

On the other hand, linear convergence is granted when the lower eigenvalue of $\mathbb{H}_{\mathcal{L}}$ is strictly positive (Peyré 2020), because the Hessian's condition number $\kappa(\mathbb{H}_{\mathcal{L}}) = \frac{\lambda_{\max}}{\lambda_{\min}}$ is bounded in the interval $\left[1, \frac{\beta}{\alpha}\right]$. When $\kappa\left(\mathbb{A}^T\mathbb{A}\right)$ is large, the contraction constant $\left(\frac{\frac{\beta}{\alpha}-1}{\frac{\beta}{\alpha}+1}\right)^2$ in equation [5.83] scales like $1 - \frac{2\alpha}{\beta}$. Faster convergence is achieved for smaller condition number, that is, for a Hessian with $\lambda_{\max} \approx \lambda_{\min}$ and $\mathbb{H}_{\mathcal{L}} \approx \lambda\mathbb{I}$. More in general, if the Hessian is bounded by β ($\mathbb{H}_{\mathcal{L}} \preceq \beta\mathbb{I}$), but positive semi-definite ($\mathbb{H}_{\mathcal{L}} \succeq 0$), some of its eigenvalues can be 0 and therefore its condition number can go to infinite. The minimum is not unique, with ker$(\mathbb{A}) \not\subset \{\mathbf{0}\}$. For such *stiff* problems, the linear convergence of the *delta rule* in equation [5.83] is not granted, but only a sub-linear convergence is obtained (Peyré 2020). The importance of the positive definiteness of the Hessian is also highlighted by the fact that if $\mathbb{H}_{\mathcal{L}} \succeq \alpha\mathbb{I}$, the quadratic loss function is *coercive*[28], because it exists $\alpha \in \mathbb{R}_*^+$ such that $\|\mathbb{A}\boldsymbol{\theta}\|^2 \geq \alpha\|\boldsymbol{\theta}\|^2$. Coercivity implies positive definiteness of the Hessian, $\ker(\mathbb{A}) = \{\mathbf{0}\}$ and therefore a unique global minimizer of the quadratic loss function on a convex set $\Theta \subset \mathbb{R}^{d_\Theta}$. In this case, $\alpha\mathbb{I} \preceq \mathbb{H}_\mathcal{L} \preceq \beta\mathbb{I}$ and all the eigenvalues of the Hessian are positive and belonging to $[\alpha, \beta]$.

When the Hessian of a quadratic function is positive definite, $\mathbb{H}_\mathcal{L} \succ 0$, the loss function is strictly convex:

$$\left\langle \boldsymbol{\nabla}_{\boldsymbol{\theta}}\mathcal{L}(\boldsymbol{\theta} + \boldsymbol{\delta\theta}) - \boldsymbol{\nabla}_{\boldsymbol{\theta}}\mathcal{L}(\boldsymbol{\theta}), \boldsymbol{\delta\theta} \right\rangle = \langle \mathbb{A}^T\mathbb{A}\boldsymbol{\delta\theta}, \boldsymbol{\delta\theta}\rangle$$
$$= \langle \mathbb{H}_\mathcal{L}\boldsymbol{\delta\theta}, \boldsymbol{\delta\theta}\rangle > 0$$

The converse, however, is not true: strictly convex functions have positive semi-definite Hessians. This statement is important for non-quadratic functions that are at least $\mathcal{C}^1(\Theta)$ and twice differentiable in a neighborhood of a local minimizer $\hat{\boldsymbol{\theta}}$. In this case, their second-order Taylor's expansion in such a neighborhood reads:

$$\mathcal{L}\left(\hat{\boldsymbol{\theta}} + \boldsymbol{\delta\theta}\right) = \mathcal{L}\left(\hat{\boldsymbol{\theta}}\right) + \left\langle \boldsymbol{\nabla}_{\boldsymbol{\theta}}\mathcal{L}\left(\hat{\boldsymbol{\theta}}\right), \boldsymbol{\delta\theta} \right\rangle + \frac{1}{2}\left\langle \mathbb{H}_\mathcal{L}\left(\hat{\boldsymbol{\theta}}\right).\boldsymbol{\delta\theta}, \boldsymbol{\delta\theta} \right\rangle + o(\|\boldsymbol{\delta\theta}\|^2) \qquad [5.85]$$

Provided the Peano–Kantorovitch inequality ($\left\langle \boldsymbol{\nabla}_{\boldsymbol{\theta}}\mathcal{L}\left(\hat{\boldsymbol{\theta}}\right), \boldsymbol{\delta\theta} \right\rangle \geq 0$ for all $\boldsymbol{\delta\theta}$), equation [5.85] grants that $\mathcal{L}(\hat{\boldsymbol{\theta}} + \boldsymbol{\delta\theta}) > \mathcal{L}\left(\hat{\boldsymbol{\theta}}\right)$ for any small increment $\boldsymbol{\delta\theta}$, if and only if the symmetric bilinear form represented by the second-order term $\left\langle \mathbb{H}_\mathcal{L}\left(\hat{\boldsymbol{\theta}}\right).\boldsymbol{\delta\theta}, \boldsymbol{\delta\theta} \right\rangle > 0$, that is, for positive definite Hessian ($\mathbb{H}_\mathcal{L}(\boldsymbol{\theta}) \succ 0$). In other words, local positive definiteness of the Hessian grants the local convergence of the *delta rule* toward a minimizer.

28 $\lim_{\|\boldsymbol{\theta}\| \to +\infty} \mathcal{L}(\boldsymbol{\theta}) = +\infty$.

Coercive quadratic functions are characterized by the condition $\mathbb{H}_{\mathcal{L}} \succeq \alpha \mathbb{I}$, which implies their α-strong convexity[29], as proved by the following expression:

$$\left\langle \nabla_{\boldsymbol{\theta}} \mathcal{L}(\boldsymbol{\theta}+\delta\boldsymbol{\theta}) - \nabla_{\boldsymbol{\theta}} \mathcal{L}(\boldsymbol{\theta}), \delta\boldsymbol{\theta} \right\rangle$$
$$= \langle \mathbb{A}^T \mathbb{A} \delta\boldsymbol{\theta}, \delta\boldsymbol{\theta} \rangle = \langle \mathbb{H}_{\mathcal{L}} \delta\boldsymbol{\theta}, \delta\boldsymbol{\theta} \rangle \geq \alpha \|\delta\boldsymbol{\theta}\|^2$$

In such as circumstance, the quadratic loss function belongs to the class of *elliptic functions*, that is, $\mathcal{C}^1(\Theta)$ functions that are α-strongly convex. From a geometric standpoint, elliptic functions have larger curvature than a α-coercive quadratic form, which represents its lower quadratic tangential minorant. Using the second-order Taylor's expansion of $\mathcal{L}(\boldsymbol{\theta}+\delta\boldsymbol{\theta})$ with an integral reminder, with $\boldsymbol{\theta}_t = \boldsymbol{\theta} + t\delta\boldsymbol{\theta}$, $t \in [0, 1]$, we obtain that:

$$\mathcal{L}(\boldsymbol{\theta}+\delta\boldsymbol{\theta}) - \mathcal{L}(\boldsymbol{\theta}) = \left\langle \nabla_{\boldsymbol{\theta}} \mathcal{L}(\boldsymbol{\theta}), \delta\boldsymbol{\theta} \right\rangle + \int_0^1 \left\langle \nabla_{\boldsymbol{\theta}} \mathcal{L}(\boldsymbol{\theta}_t) - \nabla_{\boldsymbol{\theta}} \mathcal{L}(\boldsymbol{\theta}), \delta\boldsymbol{\theta} \right\rangle dt$$

Considering the inequality for α-strongly convex functions, for which $\left\langle \nabla_{\boldsymbol{\theta}} \mathcal{L}(\boldsymbol{\theta}+\delta\boldsymbol{\theta}) - \nabla_{\boldsymbol{\theta}} \mathcal{L}(\boldsymbol{\theta}), \delta\boldsymbol{\theta} \right\rangle \geq \alpha \|\delta\boldsymbol{\theta}\|^2$, and the fact that

29 α-Strongly convex functions $\mathcal{L} : \Theta \subset \mathbb{R}^{d_\Theta} \to \mathbb{R}$, defined over a convex set Θ, satisfy the following inequality by definition:

$$\forall (\boldsymbol{\theta}_1, \boldsymbol{\theta}_2) \in \Theta^2, \forall t \in [0,1]\ \mathcal{L}(\boldsymbol{\theta}_t)$$
$$= \mathcal{L}(t\boldsymbol{\theta}_2 + (1-t)\boldsymbol{\theta}_1) \leq t \cdot \mathcal{L}(\boldsymbol{\theta}_2) + (1-t) \cdot \mathcal{L}(\boldsymbol{\theta}_1)$$
$$-\alpha t(1-t)\|\boldsymbol{\theta}_2 - \boldsymbol{\theta}_1\|^2$$

$\alpha > 0$ The previous condition on α-strong convexity is equivalent, for functions that are $\mathcal{C}^1(\Theta)$, to the inequality:

$$\mathcal{L}(\boldsymbol{\theta}_2) \geq \mathcal{L}(\boldsymbol{\theta}_1) + \langle \nabla_{\boldsymbol{\theta}} \mathcal{L}(\boldsymbol{\theta}_1), \boldsymbol{\theta}_2 - \boldsymbol{\theta}_1 \rangle + \frac{\alpha}{2} \|\boldsymbol{\theta}_2 - \boldsymbol{\theta}_1\|^2$$

$\forall (\boldsymbol{\theta}_1, \boldsymbol{\theta}_2) \in \Theta^2$ and to other inequality that reads:

$$\langle \nabla_{\boldsymbol{\theta}} \mathcal{L}(\boldsymbol{\theta}_2) - \nabla_{\boldsymbol{\theta}} \mathcal{L}(\boldsymbol{\theta}_1), \boldsymbol{\theta}_2 - \boldsymbol{\theta}_1 \rangle \geq \alpha \|\boldsymbol{\theta}_2 - \boldsymbol{\theta}_1\|^2 \qquad \forall (\boldsymbol{\theta}_1, \boldsymbol{\theta}_2) \in \Theta^2$$

$\boldsymbol{\delta\theta} = \frac{\boldsymbol{\theta}_t - \boldsymbol{\theta}}{t}$ if $t \neq 0$ and $\lim_{t \to 0} \frac{\boldsymbol{\theta}_t - \boldsymbol{\theta}}{t} = \boldsymbol{\delta\theta}$, the Taylor's expansion with integral reminder is bounded by below according to the following expression:

$$\mathcal{L}(\boldsymbol{\theta} + \boldsymbol{\delta\theta}) - \mathcal{L}(\boldsymbol{\theta}) \geq \left\langle \boldsymbol{\nabla}_{\boldsymbol{\theta}} \mathcal{L}(\boldsymbol{\theta}), \boldsymbol{\delta\theta} \right\rangle + \alpha \int_0^1 t \|\boldsymbol{\delta\theta}\|^2 dt$$

Elliptic functions with β-Lipschitz gradient are bounded by above and by below by quadratic tangential majorant and minorant, respectively (Peyré 2020).

PROPOSITION 5.1.– A strongly convex function $\mathcal{L} : \Theta \to \mathbb{R}$, defined over a non-empty convex set $\Theta \subset \mathbb{R}^{d_\Theta}$, $\mathcal{L}$ proper and strongly convex on Θ, with $\mathcal{L} \in \mathcal{C}^1(\Theta)$ and a gradient $\boldsymbol{\nabla}_{\boldsymbol{\theta}} \mathcal{L}$ β-Lipschitz on Θ, then $\forall \boldsymbol{\theta} \in \Theta$:

$$\begin{aligned} &\frac{\alpha}{2} \|\boldsymbol{\delta\theta}\|^2 \leq \mathcal{L}(\boldsymbol{\theta} + \boldsymbol{\delta\theta}) \\ &- \mathcal{L}(\boldsymbol{\theta}) - \left\langle \boldsymbol{\nabla}_{\boldsymbol{\theta}} \mathcal{L}, \boldsymbol{\delta\theta} \right\rangle \leq \frac{\beta}{2} \|\boldsymbol{\delta\theta}\|^2, \quad \forall \boldsymbol{\delta\theta} \end{aligned} \qquad [5.86]$$

If $\mathcal{L} \in \mathcal{C}^2$:

$$\alpha \mathbb{I} \preceq \mathbb{H}_{\mathcal{L}}(\boldsymbol{\theta}) \preceq \beta \mathbb{I}, \qquad \forall \boldsymbol{\theta} \in \Theta \qquad [5.87]$$

If the loss function is α-strongly convex and sufficiently regular to have a β-Lipschitz gradient, theorem 5.8 can be extended as follows.

THEOREM 5.9.– Provided the same assumptions of theorem 5.8, if $\mathcal{L}$ is α-strongly convex, there exists $0 \leq \rho(\alpha) < 1$ such that:

$$\|\boldsymbol{\theta}^{(i)} - \hat{\boldsymbol{\theta}}_N\| \leq \rho^i(\alpha) \|\boldsymbol{\theta}^{(0)} - \hat{\boldsymbol{\theta}}_N\| \qquad [5.88]$$

equation [5.88] can be rewritten as (Nesterov 1983; Peyré 2020):

$$\|\boldsymbol{\theta}^{(n_e)} - \hat{\boldsymbol{\theta}}_N\| \leq \left(\frac{\kappa(\mathbb{H}_{\mathcal{L}}(\boldsymbol{\theta}))}{1 + \kappa(\mathbb{H}_{\mathcal{L}}(\boldsymbol{\theta}))} \right)^{\frac{n_e}{2}} \|\boldsymbol{\theta}^{(0)} - \hat{\boldsymbol{\theta}}_N\| \qquad [5.89]$$

Intuitively, the condition number of the Hessian measures the "skewness" of the loss function with respect to the minorant and majorant tangent coercive quadratic functions of constant $|\lambda_{\min}(\mathbb{H}_{\mathcal{L}})|$ and $|\lambda_{\max}(\mathbb{H}_{\mathcal{L}})|$ (Murphy 2022). From a statistical standpoint, close to the MLE $\hat{\boldsymbol{\theta}}_N$, the Hessian of the negative log-likelihood (the loss) corresponds to the average observed FIM (see equation [5.35]) and to the negative conditional covariance of the observations, in case of exponential families (see equation [5.55]). Well-conditioned observed FIM corresponds to a larger Fisher information, that is, to a sharper minimum and lower variance of the MLE (lower CRB, see equation [5.38]). In other words, the better conditioned the Hessian, the lower will be the uncertainty on the MLE estimator.

Based on this statistical interpretation, the Hessian of the loss function is widely used in second-order gradient descent methods, which converge faster than first-order methods, but demand a cumbersome operation of inversion of the Hessian. Second-order methods sprout from the original *Newton's method*, whose iterative update rule reads:

$$\boldsymbol{\theta}^{(i+1)} = \boldsymbol{\theta}^{(i)} - \mathbb{H}_{\mathcal{L}}^{-1}\left(\boldsymbol{\theta}^{(i)}\right).\boldsymbol{\nabla}_{\boldsymbol{\theta}}\mathcal{L}\left(\boldsymbol{\theta}^{(i)}\right) \qquad [5.90]$$

In analogy with equation [5.91], the Newton's method consists of minimizing the function $h\left(\boldsymbol{\delta\theta}^{(i)}\right) = \mathcal{L}\left(\boldsymbol{\theta}^{(i)} + \boldsymbol{\delta\theta}^{(i)}\right) + \frac{c}{2}\|\mathbb{H}_{\mathcal{L}}^{\frac{1}{2}}\left(\boldsymbol{\theta}^{(i)}\right)\boldsymbol{\delta\theta}^{(i)}\|^2$, which corresponds to a change of metric by means of the Hessian, since $\|\mathbb{H}_{\mathcal{L}}^{\frac{1}{2}}\left(\boldsymbol{\theta}^{(i)}\right)\boldsymbol{\delta\theta}^{(i)}\| = \|\boldsymbol{\delta\theta}^{(i)}\|_{\mathbb{H}_{\mathcal{L}}}$. The solution of the constrained minimization problem represented by finding the increment that minimizes h has the generic solutions $\boldsymbol{\nabla}_{\boldsymbol{\theta}}\mathcal{L}\left(\boldsymbol{\theta}^{(i)} + \boldsymbol{\delta\theta}^{(i)}\right) = -c\mathbb{H}_{\mathcal{L}}\left(\boldsymbol{\theta}^{(i)}\right)\boldsymbol{\delta\theta}^{(i)}$. Since the gradient is continuous by assumption, it holds that (Peyré 2020):

$$\underset{\|\boldsymbol{\delta\theta}^{(i)}\|_{\mathbb{H}_{\mathcal{L}}}=1}{\arg\min}\ \mathcal{L}\left(\boldsymbol{\theta}^{(i)} + \boldsymbol{\delta\theta}^{(i)}\right) = -\frac{\boldsymbol{\nabla}_{\boldsymbol{\theta}}\mathcal{L}\left(\boldsymbol{\theta}^{(i)}\right)}{\|\boldsymbol{\nabla}_{\boldsymbol{\theta}}\mathcal{L}\left(\boldsymbol{\theta}^{(i)}\right)\|_{\mathbb{H}_{\mathcal{L}}}} \qquad [5.91]$$

Newton's method enhances the updates of the parameters following the negative gradient direction projected onto the eigenvectors ϕ_k of the Hessian function and by scaling those updates by the corresponding eigenvalues, that is:

$$\delta\boldsymbol{\theta}^{(i)} = -\sum_{k}^{d_\Theta} \lambda_k \left(\mathbb{H}_{\mathcal{L}}\right) \left\langle \boldsymbol{\nabla}_{\boldsymbol{\theta}} \mathcal{L}\left(\boldsymbol{\theta}^{(i)}\right), \boldsymbol{\phi}_k \right\rangle \boldsymbol{\phi}_k \qquad [5.92]$$

This second-order method privileges the updates along the directions for which the observed Fisher matrix is larger (the empirical average Hessian), stretching the weight update along these directions. This corresponds to update $\boldsymbol{\theta}^{(i)}$ along the score *principal components* (see section 3.3.2), recomputed at iteration i and corresponding to the directions of largest variance of the MLE score at epoch i. Those principal components of the score are steering the quest for the MLE. The eigendirections associated with the smallest Hessian eigenvalues are less informative since the score is varying less along such directions. The first-order approximation defined by equation [5.65] is a special case of the generalized gradient descent in equation [5.90], for $\mathbb{H}_{\mathcal{L}}^{-1}\left(\boldsymbol{\theta}^{(i)}\right) = \eta^{(i)}\mathbb{I}$.

Second-order methods converge faster than the first method, thanks to the projection of the negative gradient along the directions of maximum variance of the score. However, computing and inverting the Hessian at each iteration is computationally prohibitive. The most widely known alternative to Newton's method is the BFGS (Broyden–Fletcher–Goldfarb–Shanno) algorithm as well as its limited-memory version (L-BFGS); it is a quasi-Newton method that attempts to approximate the inverse of the Hessian, without any matrix inversion. One iteration of the BFGS algorithm has a computational complexity $\mathcal{O}(d_\Theta^2)$, compared to standard Newton's method (complexity $\mathcal{O}(d_\Theta^3)$) (Peyré 2020).

Newton's method and BFGS or other quasi-Newton methods are rarely adopted in machine learning practice, because their computational cost hinders efficient training on standard GPUs, especially when d_Θ is large. Not only the Hessian must be computed and inverted at each iteration, but also stored, causing intensive memory usage on CPU. First-order gradient methods are the common standards instead and a non-exhaustive list will be presented in the reminder of the section, before a small digression on the use of BatchNorm.

5.3.3. *The role of BatchNorm*

The optimal conditioning is achieved for the Hessian that is close to the identity. In the supervised case, for *regular* statistical models (see remark 5.5 for the definition), the average Hessian corresponds to the observed FIM, defined in equation [5.34]. Moreover, if $p_{\boldsymbol{\theta}}$ belongs to the family of exponential probability distributions maximizing the conditional empirical entropy $H\left(p_{\boldsymbol{\theta}}\left(\boldsymbol{Y}|\boldsymbol{X}\right)\right)$ (according to the principle of maximum entropy outlined in section 5.2.5), it can be generically expressed as:

$$p_{\boldsymbol{\theta}}\left(\boldsymbol{y}|\boldsymbol{x}\right) = \frac{e^{-\left\langle \boldsymbol{\theta}, \boldsymbol{h}_{\boldsymbol{\theta}}(\boldsymbol{x})\right\rangle}}{Z\left(\boldsymbol{\theta}|\boldsymbol{x}\right)}$$

with the equipartition function $Z\left(\boldsymbol{\theta}|\boldsymbol{x}\right) = \int_{\mathcal{Y}} e^{-\left\langle \boldsymbol{\theta}, \boldsymbol{h}_{\boldsymbol{\theta}}(\boldsymbol{x})\right\rangle} \cdot \mu\left(\boldsymbol{dy}\right)$, a normalization constant. Equation [5.55], with $\phi\left(\boldsymbol{\theta}\right) = \boldsymbol{\theta}$ states that the associated FIM corresponds to the minus covariance of the observed statistics $-\mathbb{C}_{\boldsymbol{x}\sim p}\left[\boldsymbol{h}_{\boldsymbol{\theta}}\left(\boldsymbol{X}\right)\right]$ (Campagne 2022). In this case, each diagonal term of the Hessian corresponds to the variance $\mathbb{V}_{\boldsymbol{x}\sim p}\left(h_{\boldsymbol{\theta},k}\left(\boldsymbol{X}\right)\right)$. For the sake of clarity, recalling section 5.2.4, the FIM of an MVN (belonging to the exponential family) reads:

$$\begin{aligned}\mathbb{I}^{F}\left(\boldsymbol{\theta}\right) &= \left(\boldsymbol{\nabla}_{\boldsymbol{\theta}_{\mu}} \otimes \boldsymbol{\mu}\right).\boldsymbol{\Sigma}^{-1}.\left(\boldsymbol{\mu} \otimes \boldsymbol{\nabla}_{\boldsymbol{\theta}_{\mu}}\right)\\ &+\frac{1}{2}\sum_{i,j=1}^{d_{\Theta}} \operatorname{Tr}\left(\boldsymbol{\Sigma}^{-1}\frac{\partial \boldsymbol{\Sigma}}{\partial \theta_{\boldsymbol{\Sigma},i}}\boldsymbol{\Sigma}^{-1}\frac{\partial \boldsymbol{\Sigma}}{\partial \theta_{\boldsymbol{\Sigma},j}}\right) \boldsymbol{e}_i \otimes \boldsymbol{e}_j\end{aligned}$$

and if $\boldsymbol{\mu}(\boldsymbol{\theta}_{\mu}) = \boldsymbol{\theta}_{\mu}$ and $\boldsymbol{\Sigma}(\boldsymbol{\theta}_{\Sigma}) = \boldsymbol{\theta}_{\Sigma}$, then the FIM reads:

$$\mathbb{I}^{F}\left(\boldsymbol{\theta}\right) = \boldsymbol{\Sigma}^{-1} + \frac{1}{2}\operatorname{Tr}\left(\boldsymbol{\Sigma}^{-2}\right)\mathbb{I} \qquad [5.93]$$

On a side note, if the number of samples N is sufficiently large, the MLE $\hat{\boldsymbol{\theta}}_N$ is distributed according to $\mathcal{N}\left(\mathbf{0}, \mathbb{I}_F^{-1}\right)$ (see theorem 5.3). Adopting the Newton's update in equation [5.90] requires to compute the empirical covariance of the data $\hat{\mathbb{C}}_{\boldsymbol{x}\sim p}\left[\boldsymbol{h}_{\boldsymbol{\theta}}\left(\boldsymbol{X}\right)\right] \approx \Sigma$ and invert it. However, equation [5.93] does not depend on the empirical average $\hat{\mathbb{E}}_{\boldsymbol{x}\sim p}\left[\boldsymbol{h}_{\boldsymbol{\theta}}\left(\boldsymbol{X}\right)\right] \approx \boldsymbol{\mu}$, therefore Newton's update in equation [5.90] is insensitive to the progressive bias produced by the parameter update. This justifies the empirical findings by LeCun et al. (1998) and confirmed by Wiesler et al. (2014) that any linear transformation should detrend $\boldsymbol{h}_{\boldsymbol{\theta}}$ and force the variance of its components to unit variance to be beneficial for the speed-up of the first-order gradient-descent algorithm. Not only does such a linear transformation implicitly improve the conditioning of the Hessian, but it also reveals the outputs, to avoid spurious biases. This transformation goes under the name of *batch normalization* (BatchNorm) and it consists of separately normalizing each component of $h_{\boldsymbol{\theta},k}$ and replaces it by:

$$\hat{h}_{\boldsymbol{\theta},k} = \frac{h_{\boldsymbol{\theta},k} - \hat{\mathbb{E}}_{\boldsymbol{x}\sim p}\left[h_{\boldsymbol{\theta},k}\left(\boldsymbol{X}\right)\right]}{\hat{\mathbb{V}}_{\boldsymbol{x}\sim p}\left(h_{\boldsymbol{\theta},k}\left(\boldsymbol{X}\right)\right)} \qquad [5.94]$$

with $\hat{\mathbb{E}}_{\boldsymbol{x}\sim p}\left[\cdot\right]$ and $\hat{\mathbb{V}}_{\boldsymbol{x}\sim p}\left[\cdot\right]$ being the empirical average and variance computed on the i.i.d. dataset. From another perspective, BatchNorm alleviates the so-called *internal covariance shift*, a phenomenon first identified by Shimodaira (2000), which is particularly amplified in deep *NN* with intricate architectures (as detailed in section 5.1). In such a case, the output $\boldsymbol{h}_{\boldsymbol{\theta}}\left(\boldsymbol{x}_k\right)$ is usually the result of the composition of functions, for example, for a single *hidden layer*, $\boldsymbol{h}_{\boldsymbol{\theta}}\left(\boldsymbol{x}_k\right) = \boldsymbol{h}^{(o)}_{\boldsymbol{\theta}_{(o)}} \circ \boldsymbol{h}^{(1)}_{\boldsymbol{\theta}_{(1)}}\left(\boldsymbol{x}_k\right)$. In this case, as the training epochs go by, the distribution of $\boldsymbol{\theta}_{(o)}$ changes according to the updated values of $\boldsymbol{\theta}_{(1)}$, because its update rule depends on $\boldsymbol{\theta}_{(1)}$:

$$\boldsymbol{\theta}^{(i+1)}_{(o)} = \boldsymbol{\theta}^{(i)}_{(o)} - \eta^{(i)} \boldsymbol{\nabla}_{\boldsymbol{\theta}_{(o)}} \mathcal{L}\left(\boldsymbol{\theta}^{(i)}_{(o)}, \boldsymbol{\theta}^{(i)}_{(1)}\right)$$

The *covariance shift* of $\boldsymbol{\theta}_{(o)}$ depends on the value of $\boldsymbol{\theta}_{(1)}$, whose distribution changes independently. This phenomenon is recurrent at the output of each hidden layer $\boldsymbol{h}^{(\ell)}$ if the chain of compositions of functions is longer (deeper NN, see section 5.2). Without any whitening of the output, standard first-order gradient-descent will slow down because weights follow different distributions across the architecture and first-order methods do not take into consideration the iteration-wise inverse Hessian (or observed FIM) – which privileges the updates along the directions of large score variance, by stretching the corresponding updates accordingly – nor the bias shift, which is overlooked even by applying Newton's method. Computing and inverting the Hessian is prohibitive from a computational standpoint, therefore Ioffe and Szegedy (2015) proposed a learnable version of the BatchNorm expressed in equation [5.94], by adding two extra learnable parameters β and γ according to the following expression:

$$\text{BatchNorm}\left(h_{\boldsymbol{\theta},k}\right) = \frac{h_{\boldsymbol{\theta},k} - \hat{\mathbb{E}}_{\boldsymbol{x}\sim p_{\boldsymbol{\theta}}}\left[h_{\boldsymbol{\theta},k}\left(\boldsymbol{X}\right)\right]}{\hat{\mathbb{V}}_{\boldsymbol{x}\sim p_{\boldsymbol{\theta}}}\left(h_{\boldsymbol{\theta},k}\left(\boldsymbol{X}\right)\right)} \cdot \beta + \gamma \quad [5.95]$$

$\hat{\mathbb{E}}_{\boldsymbol{x}\sim p_{\boldsymbol{\theta}}}\left[h_{\boldsymbol{\theta},k}\left(\boldsymbol{X}\right)\right]$ and $\hat{\mathbb{V}}_{\boldsymbol{x}\sim p_{\boldsymbol{\theta}}}\left(h_{\boldsymbol{\theta},k}\left(\boldsymbol{X}\right)\right)$ are computed empirically and BatchNorm is applied at the output of each intermediate layer in deep NN, adding β and γ to the set of learnable parameters $\boldsymbol{\theta}$. Since the full whitening is costly, Ioffe and Szegedy (2015) proposed the simplified version in equation [5.95], which instead of whitening the features in layer inputs and outputs jointly, each scalar feature $h_{\boldsymbol{\theta},k}$ is normalized independently. This learnable approach preserves the identity, when needed, by letting $\beta = \hat{\mathbb{E}}_{\boldsymbol{x}\sim p_{\boldsymbol{\theta}}}\left[h_{\boldsymbol{\theta},k}\left(\boldsymbol{X}\right)\right]$ and $\gamma = \hat{\mathbb{V}}_{\boldsymbol{x}\sim p_{\boldsymbol{\theta}}}\left(h_{\boldsymbol{\theta},k}\left(\boldsymbol{X}\right)\right)$. BatchNorm in equation [5.95] is beneficial but not always sufficient because without computing them explicitly (because of the computational cost implied), the eigendirections of the Hessian are a priori unknown and generally do not coincide with the direction of $\boldsymbol{h}_{\boldsymbol{\theta}}$. To see this, one can consider the FIM of an MVN with $\boldsymbol{\mu}(\boldsymbol{\theta}_{\boldsymbol{\mu}}) = \boldsymbol{W}\boldsymbol{\theta}_{\boldsymbol{\mu}}$, which reads:

$$\mathbb{I}^{F}\left(\boldsymbol{\theta}\right) = \boldsymbol{W}^{T}\boldsymbol{\Sigma}^{-1}\boldsymbol{W} + \frac{1}{2}\text{Tr}\left(\boldsymbol{\Sigma}^{-2}\right)\mathbb{I} \quad [5.96]$$

Applying BatchNorm as per equation [5.95], simply normalizes the diagonal of the FIM, but the term $\boldsymbol{W}^{T}\boldsymbol{\Sigma}^{-1}\boldsymbol{W}$ in equation [5.96] implies a reprojection of the $\boldsymbol{\Sigma}$ eigendirections in the basis of the trainable weights $\boldsymbol{W}$, which change at each iteration.

However, in general, equation [5.94] facilitates the optimal second-order conditioning (smaller condition number), forcing the Hessian close (but not equal) to the identity. However, equation [5.94] only detrends the bias and forces the diagonal terms of the Hessian to 1, but the Hessian would certainly have extra-diagonal terms that prevent BatchNorm from effectively achieving optimal conditioning.

5.3.4. *Stochastic gradient descent*

SGD has already been introduced in section 2.3.2. Inspired by the Fisher's approach for maximizing the log-likelihood function[30] with the unbiased estimator, a suitable minimization strategy would be to pursue the minimum via the *delta rule* in equation [5.65] via successive random gradient updates computed from uniformly sampled[31] database $\mathcal{D}_{\mathcal{X}\mathcal{Y}}$ according to the following algorithm, called SGD:

Algorithm 5.1. Stochastic gradient descent and batch gradient descent

$i = 0$
Initialize $\boldsymbol{\theta}^{(0)}$ and $\eta^{(0)}$
while $i < n_e$ **do**
  $\eta^{(i)}$:= Schedule($\eta^{(i)}$, i, i_0, η_d $\eta_{\max}$, $\eta_{\min}$)
  for $t \sim \mathcal{U}\left(\{k\}_{k=1}^{N}\right)$ **do**
    $\boldsymbol{\theta}^{(i+1)} = \boldsymbol{\theta}^{(i)} - \eta^{(i)} \boldsymbol{\nabla}_{\boldsymbol{\theta}} \mathcal{L}\left(\boldsymbol{h}_{\boldsymbol{\theta}}\left(\boldsymbol{x}_t\right), \boldsymbol{y}_t; \boldsymbol{\theta}^{(i)}\right), \quad \eta^{(i)} \in \mathbb{R}^+$
  end for
end while

30 Minimizing the negative log-likelihood that belongs to the family of empirical loss functions.

31 The uniformly random sampling is performed on the indices of the initial database. In the algorithm box, the uniform distribution is indicated by $\mathcal{U}$.

In other words, SGD approximates the gradient in equation [5.63] at each epoch, by randomly sampling a datum and updating the gradient upfront. The *delta rule* is applied N times per epoch. Statistically speaking, $\boldsymbol{\theta}^{(i+1)}$ are random vectors that converge toward $\hat{\boldsymbol{\theta}}_N$ at a certain speed. SGD effectively reduces the computational burden of the original GD, equal to $O(N \cdot d_\Theta)$, to $\mathcal{O}(d_\Theta)$.

Within the framework of the SGD algorithm, the *delta rule* in equation [5.65] is usually applied by averaging the randomly sampled gradients on "mini-batches" (see section 2.3.3 for an example) of N_b instances each, that is, adopting the following algorithm:

Algorithm 5.2. Stochastic gradient descent on mini-batches

$i = 0$
Initialize $\boldsymbol{\theta}^{(0)}$ and $\eta^{(0)}$
while $i < n_e$ **do**
    $\eta^{(i)}$:= Schedule($\eta^{(i)}$, i, i_0, η_d $\eta_{\max}$, $\eta_{\min}$)
    for $j \sim \mathcal{U}\left(\{i\}_{i=1}^{\frac{N}{N_b}}\right)$ **do** (random mini-batch sampling)
        $\boldsymbol{g_\theta} = 0$
        for $t \sim \mathcal{U}\left(\{k_j\}_{k=1}^{N_b}\right)$ **do** (random sampling over the N_b samples within the jth batch)
            $\boldsymbol{g_\theta} += \boldsymbol{\nabla_\theta} \ell\left(\boldsymbol{h_\theta}\left(\boldsymbol{x}_t\right), \boldsymbol{y}_t; \boldsymbol{\theta}^{(i)}\right)$
        end for
        $\boldsymbol{\theta}^{(i+1)} = \boldsymbol{\theta}^{(i)} - \frac{\eta^{(i)}}{N_b} \boldsymbol{g_\theta}, \quad \eta^{(i)} \in \mathbb{R}^+$
    end for
end while

Averaging on mini-batches is certainly cost effective compared to standard SGD, which corresponds to N mini-batches of $N_b = 1$ instance each. SGD on mini-batches can be seen as a vectorized version of the SGD.

EXAMPLE 5.4 (**Stochastic gradient descent for linear regression**).–

```
import torch
import torch.nn as nn
import torch.optim as optim
import numpy as np
from matplotlib import pyplot as plt
# assure reproducibility
torch.manual_seed(0)
# Data for regression problem
A = 10.1542550
b = 14.7351129
c = 2.34
epsilon = 10.154988
N = 200 # number i.i.d. samples
Nb = 10 # number of samples in the mini-batch
# Data Generation
np.random.seed(42)
x = np.random.rand(N, 1)
y = b + A * np.exp(c*x) + epsilon * np.random.randn(N, 1)

# Shuffled indices
idx = np.arange(N)
np.random.shuffle(idx)

Ntrain = int(N*0.8)
# Uses first 80% of the dataset for training
train_idx = idx[:Ntrain]
# Uses the last 20% for validation
val_idx = idx[Ntrain:]

# Generates train and validation sets
x_train, y_train = x[train_idx], y[train_idx]
x_val, y_val = x[val_idx], y[val_idx]

# Convert Numpy arrays into PyTorch Tensors
x_train_tensor = torch.from_numpy(x_train).float()
y_train_tensor = torch.from_numpy(y_train).float()

eta = 1e-1 # constant learning rate
# for warmup: https://github.com/Tony-Y/pytorch_warmup
n_e = 1000 # number of epochs

# Creating a MLP model with 1 layer and 1 neuron
h_theta = nn.Sequential(nn.Linear(1, 1), nn.SiLU())
```

```
# Define the SGD optimizer
optimizer = optim.SGD(h_theta.parameters(), lr=eta)
L_Dxy = nn.MSELoss()

# Loop over the training epochs on training dataset
i=0
while i < n_e: # loop over epochs
  for j in range(0,int(Ntrain/Nb)): # loop over mini-batches
    t_idx = np.arange(j*Nb,(j+1)*Nb,1) # mini-batch linear
                                        # indexing
    np.random.shuffle(t_idx) # shuffle mini-batch
    x_b = x_train_tensor[t_idx]
    y_b = y_train_tensor[t_idx]
    optimizer.zero_grad() # initialize the optimizer
    loss = 0.0 # initialize the loss
    for t in range(Nb): # loop over the samples in the batch
      xt = x_b[t]
      yt = y_b[t]
      yhat = h_theta(xt) # predict the output
      loss += L_Dxy(yhat, yt) # Compute Empirical Loss
    loss.backward() # Compute derivatives with AutoGrad
    optimizer.step() # weight update!
```

5.3.5. *Beyond SGD: the role of "momentum"*

The SGD in 5.1. and its mini-batch version in 5.2. are first-order Hessian-free methods that converge slowly and are sensitive to noise (Sutskever et al. 2013; Peyré 2020). Moreover, the convergence can be hindered when the Hessian is ill-conditioned. A possible improvement is represented by the so-called "momentum" using previous gradient memory. The classical momentum technique (Polyak 1964) accelerates the gradient descent just like a "heavy ball" rolling downhill, slowing down the update in the direction with sudden gradient changes, and accelerating the convergence along the predominant historical gradient directions. In other words, it cumulates a "velocity" vector m (keeping track of the cumulated gradient across previous epochs) along the direction of a persistent reduction in $\mathcal{L}$ across iterations (Sutskever et al. 2013) and multiplies this term by a momentum coefficient $\gamma \in [0, 1)$. Algorithms 5.3. and 5.4. outline the SGD algorithm with classical momentum, without and with mini-batch average.

Algorithm 5.3. Stochastic Gradient Descent with classical momentum

$i = 0$
Initialize $\boldsymbol{\theta}^{(0)}$, $\eta^{(0)}$, $\gamma \in [0,1)$, $\tau \geq 1$
$\boldsymbol{m}^{(0)} = \mathbf{0}$
while $i < n_e$ **do**
  $\eta^{(i)}$:= Schedule($\eta^{(i)}$, i, i_0, η_d $\eta_{\max}$, $\eta_{\min}$)
  for $t \sim \mathcal{U}\left(\{k\}_{k=1}^{N}\right)$ **do**
    $\boldsymbol{m}^{(i+1)} = (1 - \langle \tau - 1 \rangle) \boldsymbol{\nabla}_{\boldsymbol{\theta}} \mathcal{L}\left(\boldsymbol{h}_{\boldsymbol{\theta}}\left(\boldsymbol{x}_t\right), \boldsymbol{y}_t; \boldsymbol{\theta}^{(i)}\right) + \gamma \boldsymbol{m}^{(i)}$
    $\boldsymbol{\theta}^{(i+1)} = \boldsymbol{\theta}^{(i)} - \eta^{(i)} \boldsymbol{m}^{(i+1)}$
  end for
end while

Algorithm 5.4. Stochastic Gradient Descent on mini-batches with classical momentum

$i = 0$
Initialize $\boldsymbol{\theta}^{(0)}$, $\eta^{(0)}$, $\gamma \in [0,1)$, $\tau \geq 1$
$\boldsymbol{m}^{(0)} = \mathbf{0}$
while $i < n_e$ **do**
  $\eta^{(i)}$:= Schedule($\eta^{(i)}$, i, i_0, η_d $\eta_{\max}$, $\eta_{\min}$)
  for $j \sim \mathcal{U}\left(\{i\}_{i=1}^{\frac{N}{N_b}}\right)$ **do** (random mini-batch sampling)
    $\boldsymbol{g}_{\boldsymbol{\theta}} = 0$
    for $t \sim \mathcal{U}\left(\{k_j\}_{k=1}^{N_b}\right)$ **do** (random sampling over the N_b samples within the jth batch)
      $\boldsymbol{g}_{\boldsymbol{\theta}} += \boldsymbol{\nabla}_{\boldsymbol{\theta}} \ell\left(\boldsymbol{h}_{\boldsymbol{\theta}}\left(\boldsymbol{x}_t\right), \boldsymbol{y}_t; \boldsymbol{\theta}^{(i)}\right)$
    end for
    $\boldsymbol{m}^{(i+1)} = (1 - \langle \tau - 1 \rangle) \boldsymbol{g}_{\boldsymbol{\theta}} + \gamma \boldsymbol{m}^{(i)}$
    $\boldsymbol{\theta}^{(i+1)} = \boldsymbol{\theta}^{(i)} - \frac{\eta^{(i)}}{N_b} \boldsymbol{m}^{(i+1)}$
  end for
end while

The factor $-\langle \tau - 1 \rangle$ dampens the gradient only if $\tau \geq 1$[32]. When no dampening is considered (i.e. $\tau \leq 1$), the classical momentum

32 $\langle \cdot \rangle$ represent the Macaulay brackets.

accelerates by a (usually) constant factor γ (often chosen equal to 0.85 or 0.95); the gradient descent in the directions $\boldsymbol{e}_\xi$ is small but persistent. For instance, in the mini-batch version in 5.4., if $\left\langle \boldsymbol{g}_{\boldsymbol{\theta}}, \boldsymbol{e}_\xi \right\rangle > 0$:

$$\left\langle \boldsymbol{e}_\xi, \boldsymbol{m}^{(i)} \right\rangle = \eta^{(i)} \left\langle \boldsymbol{e}_\xi, \boldsymbol{g}_{\boldsymbol{\theta}} \right\rangle + \gamma \left\langle \boldsymbol{e}_\xi, \boldsymbol{m}^{(i-1)} \right\rangle \geq \gamma \left\langle \boldsymbol{e}_\xi, \boldsymbol{m}^{(i-1)} \right\rangle \qquad [5.97]$$

Inertia helps to overcome local minima and saddle points. However, its effectiveness diminishes along directions where the reduction in the objective loss function is small or oscillates. These effects are amplified by algorithm 5.3. The momentum update is an equivalent of an exponentially weighted moving average of the past gradients (Murphy 2022), since $\boldsymbol{m}^{(i+1)} = \sum_{j=0}^{i} \gamma^j \boldsymbol{\nabla}_{\boldsymbol{\theta}} \mathcal{L}\left(\boldsymbol{\theta}^{(i-j)}\right)$. If the gradient updates were constant over the past iterations $\boldsymbol{m}^{(i+1)} = \boldsymbol{\nabla}_{\boldsymbol{\theta}} \mathcal{L}\left(\boldsymbol{\theta}^{(0)}\right) \sum_{j=0}^{i} \gamma^j = \frac{\boldsymbol{\nabla}_{\boldsymbol{\theta}} \mathcal{L}(\boldsymbol{\theta}^{(0)})}{1-\gamma}$, which implies that, if $\gamma = 0.9$, the initial gradient is scaled by a factor of 10, as noted by Murphy (2022).

For convex functions, momentum-based methods are known to outperform SGD in the early or transient stages (Sutskever et al. 2013). (Polyak 1964) proved that the classical momentum algorithm accelerates convergence to local minima by reducing by a factor $\sqrt{\kappa\left(\mathbb{H}_{\mathcal{L}}\right)}$ the number of iterations required by greedy algorithms. In this case, γ must be chosen equal to $\frac{\sqrt{\kappa(\mathbb{H}_{\mathcal{L}})}-1}{\sqrt{\kappa(\mathbb{H}_{\mathcal{L}})}+1}$. Second-order methods, as seen in section 5.3.2, amplify those steps in low-curvature directions, but instead of accumulating changes they re-weight the update along each eigendirection of the curvature matrix by the inverse of the associated curvature (Sutskever et al. 2013). On the contrary, classical momentum algorithms can miss the minimizer due to the consistent inertia of the gradient descent.

In `PyTorch`, the classical momentum is an option of the standard `SGD` optimizer.

EXAMPLE 5.5 (**Stochastic Gradient Descent for linear regression with classical momentum**).– This example is equivalent to example 5.4, with the exception of the definition of the optimizer.

```
...
# Define the SGD optimizer
gamma = 0.9 # classical momentum algorithm
# PyTorch adopts a classical momentum coefficient eta*mu
# which corresponds to gamma in the above-mentioned formulas
tau = 0.0 # Dampening coefficient
optimizer = optim.SGD(h_theta.parameters(), lr=eta,
        momentum=gamma, dampening=tau)
...
```

5.3.6. *Beyond classical momentum SGD: the Nesterov algorithm*

A better convergence rate than the classical momentum algorithm is provided by the Nesterov algorithm, especially for general smooth (non-strongly) convex functions and a deterministic gradient descent (Sutskever et al. 2013). The difference with the classical momentum method resides in the fact that the gradient of the loss function is computed at the position updated with $m^{(i)}$, which adds an *extrapolation step* to the original gradient descent formulation (Murphy 2022). The Nesterov accelerated gradient (NGA) algorithm reads:

Usually, γ=0.9. NGA is faster than classical momentum and even optimal for first-order gradient methods applied on convex functions with Lipschitz continuous gradient. In other cases, NGA might exhibit low convergence speed and become even unstable. Therefore, NGA must be employed with a careful choice of γ and η (Murphy 2022). In `PyTorch`, the Nesterov accelerated gradient algorithm is an option of the standard `SGD` optimizer.

EXAMPLE 5.6 (**Nesterov accelerated gradient descent for linear regression**).– This example is equivalent to example 5.4, with the exception of the definition of the optimizer.

Algorithm 5.5. Nesterov Accelerated Gradient Descent

$i = 0$
Initialize $\boldsymbol{\theta}^{(0)}, \eta^{(0)}, \gamma \in [0, 1), \tau \geq 1$
$\boldsymbol{m}^{(0)} = \mathbf{0}$
while $i < n_e$ **do**
  $\eta^{(i)} :=$ Schedule($\eta^{(i)}, i, i_0, \eta_d\ \eta_{\max}, \eta_{\min}$)
  for $t \sim \mathcal{U}\left(\{k\}_{k=1}^{N}\right)$ **do**
    $\boldsymbol{m}^{(i+1)} = \eta^{(i)}\left(1 - \langle \tau - 1 \rangle\right) \boldsymbol{\nabla}_{\boldsymbol{\theta}} \mathcal{L}\left(\boldsymbol{h}_{\boldsymbol{\theta}}\left(\boldsymbol{x}_t\right), \boldsymbol{y}_t; \boldsymbol{\theta}^{(i)} + \gamma \boldsymbol{m}_k\right) - \gamma \boldsymbol{m}^{(i)}$
    $\boldsymbol{\theta}^{(i+1)} = \boldsymbol{\theta}^{(i+1)} - \boldsymbol{m}^{(i)}$
  end for
end while

Algorithm 5.6. Nesterov Accelerated Gradient Descent on mini-batches

$i = 0$
Initialize $\boldsymbol{\theta}^{(0)}, \eta^{(0)}, \gamma \in [0, 1), \tau \geq 1$
$\boldsymbol{m}^{(0)} = \mathbf{0}$
while $i < n_e$ **do**
  $\eta^{(i)} :=$ Schedule($\eta^{(i)}, i, i_0, \eta_d\ \eta_{\max}, \eta_{\min}$)
  for $j \sim \mathcal{U}\left(\{i\}_{i=1}^{\frac{N}{N_b}}\right)$ **do** (random mini-batch sampling)
    $\boldsymbol{g}_{\boldsymbol{\theta}} = 0$
    for $t \sim \mathcal{U}\left(\{k_j\}_{k=1}^{N_b}\right)$ **do** (random sampling over the N_b samples within the jth batch)
      $\boldsymbol{g}_{\boldsymbol{\theta}} += \boldsymbol{\nabla}_{\boldsymbol{\theta}} \ell\left(\boldsymbol{h}_{\boldsymbol{\theta}}\left(\boldsymbol{x}_t\right), \boldsymbol{y}_t; \boldsymbol{\theta}_k - \gamma \boldsymbol{m}^{(i)}\right)$
    end for
    $\boldsymbol{m}^{(i+1)} = \eta^{(i)}\left(1 - \langle \tau - 1 \rangle\right) \boldsymbol{g}_{\boldsymbol{\theta}} - \gamma \boldsymbol{m}^{(i)}$
    $\boldsymbol{\theta}^{(i+1)} = \boldsymbol{\theta}^{(i)} - \frac{1}{N_b} \boldsymbol{m}^{(i+1)}$
  end for
end while

```
...
# Define the SGD optimizer
gamma = 0.9 # Nesterov momentum coefficient
tau = 0.0 # Dampening coefficient
optimizer = optim.SGD(h_theta.parameters(), lr=eta,
        nesterov=True, momentum=gamma)
...
```

5.3.7. *Optimizing with adaptive learning rates*

Several strategies have been developed to improve the convergence rate of first-order Hessian-free gradient descent algorithms. Newton's method preconditions the gradient with the full inverse of the Hessian. Quasi-Newton methods approximate the Hessian itself with a positive-definite surrogate. However, adopting such a strategy in an SGD framework, thought to efficiently run over large databases, by mini-batches of different sizes, is poorly reliable and computationally expensive: the update direction should be recomputed at each epoch, for each batch, with a full preconditioning matrix. Therefore, in practice, diagonal preconditioners are adopted and computed based on adaptive updates that take into consideration the statistics on past updates and gradients. Such statistics do not necessarily represent second-order information on the Hessian, but they speed up the standard SGD method. In the following, the most popular ones are presented.

5.3.7.1. *Adaptive gradient: AdaGrad*

(Duchi et al. 2011) proposed an adaptive update rule called *AdaGrad*. The latter is based on the definition of $\mathbb{V}^{(i)}$, defined as:

$$\mathbb{V}^{(i)} = \sum_{k=0}^{i-1} \boldsymbol{\nabla}_{\boldsymbol{\theta}} \mathcal{L}\left(\boldsymbol{\theta}^{(k)}\right) \otimes \boldsymbol{\nabla}_{\boldsymbol{\theta}} \mathcal{L}\left(\boldsymbol{\theta}^{(k)}\right) \quad [5.98]$$

The AdaGrad update rule reads:

$$\boldsymbol{\theta}^{(i+1)} = \boldsymbol{\theta}^{(i)} - \eta \left(\text{diag}\mathbb{V}^{(i)} + \epsilon\mathbb{I}\right)^{-\frac{1}{2}} . \boldsymbol{\nabla}_{\boldsymbol{\theta}} \mathcal{L}\left(\boldsymbol{\theta}^{(i)}\right) \quad [5.99]$$

with usual values of η=10^{-2} and a numeric tolerance ϵ=10^{-8}. The diagonal matrix diag($\mathbb{V}$) represents the cumulated square derivatives along θ_j, $V_{jj}^{(i)} = \sum_{k=0}^{i-1} \left(\frac{\partial \mathcal{L}}{\partial \theta_j} \left(\boldsymbol{\theta}^{(k)} \right) \right)^2$ from epoch 0 to epoch i. Moreover,

$$\sqrt{\mathrm{Tr}\left(\mathbb{V}^{(i)}\right)} = \sum_{k=0}^{i-1} \sum_{m=1}^{d_\Theta} \left(\frac{\partial \mathcal{L}}{\partial \theta_m} \left(\boldsymbol{\theta}^{(k)} \right) \right)^2$$

represents the average ℓ_2 norm of $\boldsymbol{\nabla}_{\boldsymbol{\theta}} \mathcal{L}$. Instead, after a large number of iterations i, $\frac{V_{jj}^{(i)}}{i-1}$ is an unbiased estimator of the variance $\hat{\mathbb{E}}\left[\left(\frac{\partial \mathcal{L}}{\partial \theta_j} \left(\boldsymbol{\theta}_i \right) \right)^2 \right]$ and it is monotonically increasing. AdaGrad requires to save and update diag$\mathbb{V}^{(i)}$ at each iteration. The effective learning rate of each gradient component is expressed as $\eta_j^{'} = \frac{\eta}{\sqrt{V_{jj}^{(i)} + \epsilon}}$. $\eta_j^{'}$ can become very small, due to the monotonic increase of $V_{jj}^{(i)}$, until the update stops. This aggressive learning rate decay represents one major shortcoming of the AdaGrad. To mitigate such an effect, AdaGrad update can be combined with the adaptive learning rate decay in equation [5.71].

EXAMPLE 5.7 (**AdaGrad for linear regression**).– This example is equivalent to 5.4, with the exception of the definition of the optimizer.

```
...
# Define the AdaGrad optimizer
etad = 0.0 # learning rate decay
epsilon = 1e-8 # tolerance
G0 = 0.0
optimizer = optim.Adagrad(h_theta.parameters(), lr=eta,
                          lr_decay=etad, eps=epsilon,
                          initial_accumulator_value=G0)
...
```

5.3.7.2. *Root mean square propagation: RMSprop*

Another optimization strategy, capable of mitigating the disadvantage of aggressive learning rate decay, proper to AdaGrad, was proposed by (Tieleman and Hinton 2012) and called *RMSprop*, from root mean square. The idea is to compute the exponentially weighted moving average value of $\mathbb{V}^{(i)}$, based on a shorter memory window, that is:

$$\begin{aligned}\text{diag}\tilde{\mathbb{V}}^{(i)} &= \alpha_r \,\text{diag}\tilde{\mathbb{V}}^{(i-1)} + (1-\alpha_r) \\ &\times \left(\text{diag}\mathbb{V}^{(i)} - \text{diag}\mathbb{V}^{(i-1)}\right)\end{aligned} \qquad [5.100]$$

Equation [5.100] implies the root mean square nature of RMSprop, since:

$$\sqrt{\tilde{V}_{jj}^{(i)}} = \sqrt{\frac{1}{i}\sum_{k=1}^{i} V_{jj}^{2}}$$

The update rule is the same as equation [5.99], but with $\text{diag}\tilde{\mathbb{V}}^{(i)}$ to replace $\text{diag}\mathbb{V}^{(i)}$ and usually the smoothing constant α_r= 0.9. The main difference is represented by the fact that $\text{diag}\tilde{\mathbb{V}}^{(i)}$ is not monotonically increasing, therefore the learning rate decay is slower.

EXAMPLE 5.8 (**RMSprop for linear regression**).– This example is equivalent to 5.4, with the exception of the definition of the optimizer.

```
...
# Define the RMSprop optimizer
alphar = 0.9 # smoothing constant (called alpha in PyTorch)
epsilon = 1e-8 # tolerance
GO = 0.0
optimizer = optim.RMSprop(h_theta.parameters(), lr=eta,
                          alpha=alphar, eps=epsilon)
...
```

5.3.7.3. *RMSprop with momentum: ADADELTA*

The RMSprop with momentum (see section 5.3.5) is called *ADADELTA* (Zeiler 2012). In particular, after having computed $\tilde{\mathbb{V}}^{(i)}$ defined in equation [5.100], ADADELTA computes the vector $\boldsymbol{U}^{(i+1)}$ with momentum, that is, as follows:

$$\boldsymbol{U}^{(i+1)} = \alpha_u \boldsymbol{U}^{(i)} + (1-\alpha_u)\,\boldsymbol{u}^{(i)} \qquad [5.101]$$

with $\boldsymbol{u}^{(i)}$ that slightly modifies the gradient with a term approximating the inverse Hessian matrix (diagonal), but based on the previous gradient update, according to the expression:

$$\boldsymbol{u}^{(i+1)} = \text{diag}^{\frac{1}{2}}\left(\mathbb{U}^{(i)} + \epsilon\mathbb{I}\right)\text{diag}^{-\frac{1}{2}}\left(\tilde{\mathbb{V}}^{(i)} + \epsilon\mathbb{I}\right).\boldsymbol{\nabla}_{\boldsymbol{\theta}}\mathcal{L}\left(\boldsymbol{\theta}^{(i)}\right) \quad [5.102]$$

where $\mathbb{U}^{(i)} = \left(\boldsymbol{U}^{(i)} \otimes \boldsymbol{U}^{(i)}\right)^{\frac{1}{2}}$. All the vectors $\boldsymbol{u}^{(i+1)}$ have the same dimensions of $\boldsymbol{\theta}$. The term $\text{diag}^{\frac{1}{2}}\left(\mathbb{U}^{(i)} + \epsilon\mathbb{I}\right)$ represents the inverse Hessian approximation, with the contribution of the RMSprop update, whereas the term $\text{diag}^{-\frac{1}{2}}\left(\tilde{\mathbb{V}}^{(i)} + \epsilon\mathbb{I}\right)$ the contribution of the RMSprop gradient. Finally, $\boldsymbol{\theta}^{(i+1)} = \boldsymbol{\theta}^{(i)} - \boldsymbol{u}^{(i)}$. The advantage of the ADADELTA is that it is parameter free, since no learning rate *stricto sensu* is required. This prevents the possibility of a learning rate scheduler. Another ADADELTA's shortcoming is represented by the biased momentum estimate in equations [5.101] and [5.102], strictly depending on the weights' initialization.

EXAMPLE 5.9 (**Adadelta for linear regression**).– This example is equivalent to 5.4, with the exception of the definition of the optimizer.

```
...
# Define the ADADELTA optimizer
alphau = 0.9 # smoothing constant (called rho in PyTorch)
epsilon = 1e-8 # tolerance
optimizer = optim.Adadelta(h_theta.parameters(), lr=eta,
                           rho=alphau, eps=epsilon)
...
```

5.3.7.4. *Adaptive moment estimation: Adam*

The adaptive moment estimation, *Adam*, invented by (Kingma and Ba 2015), is a rather famous optimizer in deep learning. Adam retrieves the idea of momentum, by estimating the first (mean) and second order (variance) moments. Adam endorses the idea of momentum by keeping a moving average of previous gradients (first moment), which will be more or less important depending on the value of a coefficient β_1. Adam is quite similar to ADADELTA (see section 5.3.7.3), but it adds another value (second moment), which will be the sliding average of the previous squared gradients whose square root corresponds to the sliding uncentered variance of the previous gradients. Adam computes the update according to the following steps:

– It estimates the diagonal of Hessian as done in ADADELTA by the term $\mathrm{diag}^{\frac{1}{2}}\left(\mathbb{U}^{(i)}+\epsilon\mathbb{I}\right)\mathrm{diag}^{-\frac{1}{2}}\left(\tilde{\mathbb{V}}^{(i)}+\epsilon\mathbb{I}\right)$ but with local accumulation (the decay of past square gradients):

$$\mathrm{diag}\tilde{\mathbb{V}}^{(i)}=\beta_2\tilde{\mathbb{V}}^{(i-1)}+(1-\beta_2)\left(\mathrm{diag}\mathbb{V}^{(i)}-\mathrm{diag}\mathbb{V}^{(i-1)}\right) \qquad [5.103]$$

– It estimates the first-order gradient moment, similarly to $\boldsymbol{u}^{(i+1)}$ in equation [5.101] (the weighted decay of the past gradients, preconditioned by $\mathrm{diag}^{\frac{1}{2}}\left(\mathbb{U}^{(i)}+\epsilon\mathbb{I}\right)\mathrm{diag}^{-\frac{1}{2}}\left(\tilde{\mathbb{V}}^{(i)}+\epsilon\mathbb{I}\right)$):

$$\boldsymbol{m}^{(i)}=\beta_1\boldsymbol{m}^{(i-1)}+(1-\beta_1)\,\boldsymbol{\nabla}_{\boldsymbol{\theta}}\mathcal{L}\left(\boldsymbol{\theta}^{(i)}\right) \qquad [5.104]$$

– It normalizes both $\mathrm{diag}\tilde{\mathbb{V}}^{(i)}$ and $\boldsymbol{m}^{(i)}$ to avoid producing biased estimators:

$$\hat{\boldsymbol{m}}^{(i)}=\frac{\boldsymbol{m}^{(i)}}{1-\beta_1},\qquad \hat{\mathbb{V}}^{(i)}=\frac{\tilde{\mathbb{V}}^{(i)}}{1-\beta_2} \qquad [5.105]$$

This additional operation reduces the biases of the first iterations where the moving averages are derived from a small number of values.

– It updates the weights as:

$$\boldsymbol{\theta}^{(i+1)}=\boldsymbol{\theta}^{(i)}-\eta\cdot\left(\mathrm{diag}^{\frac{1}{2}}\left(\hat{\mathbb{V}}^{(i)}\right)+\epsilon\mathbb{I}\right)^{-1}\hat{\boldsymbol{m}}^{(i)} \qquad [5.106]$$

Adam has several advantages: the per-dimension update inherited from ADADELTA, which is very important to promote and tackle sparsity. As RMSprop and ADADELTA, it adopts the use of second-order momentum on the denominator. This aspect is tightly linked to the information of the statistical model represented by *NN* parameterized by its weights and biases $\boldsymbol{\theta}$. The CRB presented in equation [5.39] provides an insightful view since it states that the Fisher information $\hat{\mathbb{I}}^F\left(\boldsymbol{\theta}\right)$ is larger the inverse variance of the weights (represented by $\mathrm{diag}^{-\frac{1}{2}}\hat{\mathbb{V}}^{(i)}$). In other words, the optimizer tries to increase the overall informative power of the *NN* to make it more flexible to complex regression or classification tasks. Moreover, Adam has the advantage of integrating momentum, but also another parameter that will prevent the momentum from taking too much importance in updating the weights, so that Adam dampens the effects of oscillation in the gradient descent.

EXAMPLE 5.10 (**Adam for linear regression**).– This example is equivalent to 5.4, with the exception of the definition of the optimizer.

```
...
# Define the SGD optimizer
beta1 = 0.9 # Adam coefficients
beta2 = 0.999 # Adam coefficients
epsilon = 1e-8 # tolerance
optimizer = optim.Adam(h_theta.parameters(), lr=eta,
                       betas=(beta1, beta2), eps=epsilon)
...
```

5.3.7.5. *Weight decay*

The weight decay allows to keep $||\boldsymbol{\theta}||$ small. Heuristically, this is a preferable condition to achieve convergence in the optimization process. Compared to explicitly penalizing the ℓ^2 norm in the loss function, by adding something of the sort of $\Omega\left(\boldsymbol{\theta}\right) = ||\boldsymbol{\theta}||_2^2$, the weight decays makes the weights tend to zero quite faster. The weight decay consists of simply adding an extra term $\lambda\boldsymbol{\omega}^{(i)}$ either to the gradient $\nabla_{\boldsymbol{\theta}}\mathcal{L}\left(\boldsymbol{\theta}^{(i)}\right)$ or to the weight update. In this latter case, the weight decay is decoupled from first- and second-order moments. For instance, with the Adam optimizer, the decoupled weight decay reads:

$$\boldsymbol{\theta}^{(i+1)} = \boldsymbol{\theta}^{(i)} - \eta \cdot \mathrm{diag}^{-\frac{1}{2}}\left(\hat{\mathbb{V}}^{(i)} + \epsilon\mathbb{I}\right)\hat{\boldsymbol{m}}^{(i)} + \lambda\boldsymbol{\omega}^{(i+1)} \quad [5.107]$$

The `PyTorch` `optimizer` class implements a `weight_decay` option. This extra parameter can be directly added to the `SGD` optimizer input list since the `SGD` does not estimate either the first or the second gradient moments. On the contrary, `RMSprop`, `Adagrad`, `AdaDelta` and `Adam` do. However, adding the `weight_decay` parameter to those standard optimizer classes, couples the weight decay with the estimation of the first and second moments, which is not the most effective way to perform weight decay, since it can add spurious biases. Instead, to apply a decoupled weight decay as in equation [5.107], one should use the `SGDW` or `AdamW` (Loshchilov and Hutter 2019) optimizers[33].

EXAMPLE 5.11 (**Compare optimizers in** `PyTorch`).– In the following example (Vialle 2022), the `SGD` optimizer, with and without (Nesterov) momentum, is compared to `Adam`, `Adadelta`, `RMSProp` and `AdamW`, in the framework of an optimization problem with $\boldsymbol{\theta} \in \mathbb{R}^2$ and with the following loss function:

$$\mathcal{L}\left(\theta_1, \theta_2\right) = 8 \cdot \sin\left(\frac{1.5 \cdot \theta_1}{10} - 3\right) \cdot \sin\left(\frac{1.5 \cdot \theta_1}{10} - 3\right) + \\ + 8 \cdot \cos\left(\frac{\theta_2}{10} - 2\right) \cdot \cos\left(\frac{\theta_2}{10} - 2\right) \\ + 0.3 \cdot \left(\left(\frac{\theta_1}{5} - 12\right) \cdot \left(\frac{\theta_1}{5} - 12\right) + \left(\frac{\theta_2}{10} - 8\right) \cdot \left(\frac{\theta_2}{10} - 8\right)\right) + 10 \qquad [5.108]$$

$\mathcal{L}\left(\boldsymbol{\theta}\right)$ is depicted in Figure 5.7.

The comparison is performed according to the following criteria:

1) parameter initialization $(\theta_1^{(0)}, \theta_2^{(0)})$ (see section 5.2.4.1 for further insights);

2) use of the (Nesterov) momentum and dampening (for `SGD`) and `RMSprop`;

3) weight decay for Adam and AdamW.

All tested optimizers share the same (constant) learning rate $\eta = 20$ and 10,000 epochs.

First and foremost, the effect of two different initializations $\boldsymbol{\theta}^{(0-1)}$=(135,135) and $\boldsymbol{\theta}^{(0-2)}$=(45,35) is tested. The choice of $\boldsymbol{\theta}^{(0-1)}$ implies that the optimizer algorithm starts from a point closer to the

33 Available at: https://pytorch.org/docs/stable/generated/torch.optim.AdamW.html

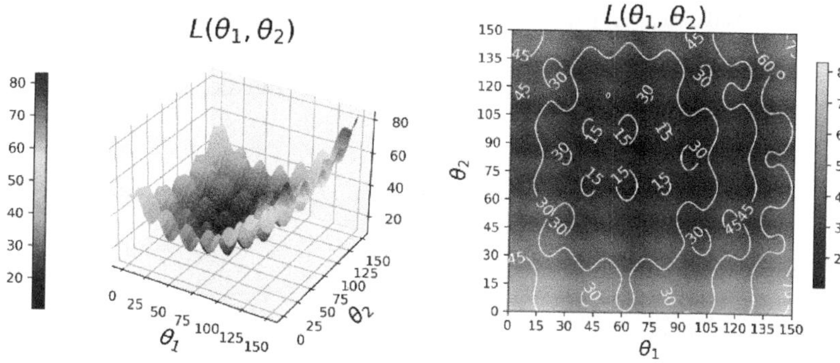

(a) (b)

Figure 5.7. *$\mathcal{L}(\boldsymbol{\theta}) : \mathbb{R}^2 \to \mathbb{R}$ defined in equation [5.108]. (a) $\mathcal{L}(\boldsymbol{\theta})$ in three dimensions. (b) Two-dimensional contour plot*

minimum $\hat{\boldsymbol{\theta}}_N$ than $\boldsymbol{\theta}^{(0-2)}$ (see Figure 5.7). Figure 5.8 shows the SGD's high sensitivity to the initialization: the algorithm diverges from $\hat{\boldsymbol{\theta}}_N$ (within the 10,000 epochs) when starting from $\boldsymbol{\theta}^{(0-2)}$ (see Figure 5.8(b)), which is closer to the minimum itself. This is mostly because the function is not convex close to the minimum $\hat{\boldsymbol{\theta}}_N$.

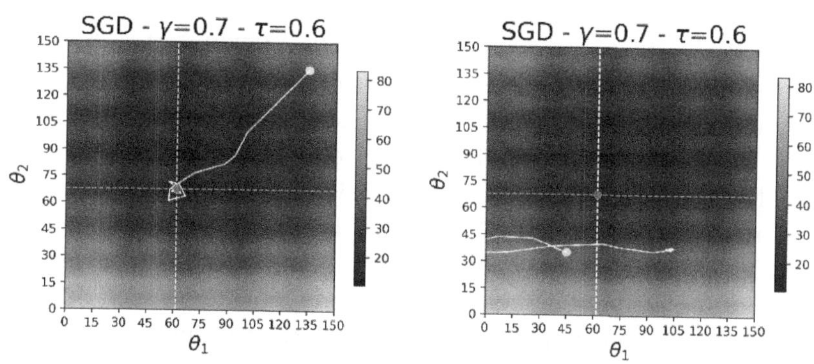

(a) (b)

Figure 5.8. *SGD minimization trajectory for $\boldsymbol{\theta}^{(0-1)}$ (a) and $\boldsymbol{\theta}^{(0-2)}$ (b). SGD is featured by momentum $\gamma = 0.7$ and dampening $\tau = 0.6$. The yellow dot represents the starting point $\boldsymbol{\theta}^{(0)}$, whereas the blue dot represents the sought minimum*

`Adadelta` (with $\alpha_u \in [0, 1]$, called `rho` in `PyTorch`) fails for both initialization choices, as depicted in Figure 5.9. This is due to its high dependency on the weight initialization with a biased momentum estimate.

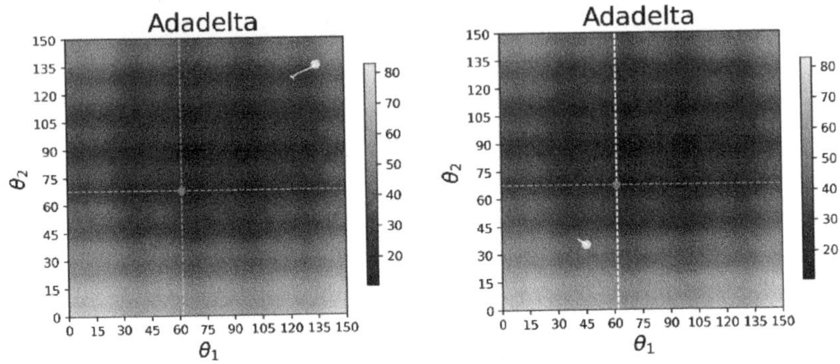

(a) (b)

Figure 5.9. *`Adadelta` minimization trajectory for $\boldsymbol{\theta}^{(0-1)}$ (a) and $\boldsymbol{\theta}^{(0-2)}$ (b). `Adadelta` is featured by α_u = 0.9, but similar results are obtained regardless of its value $\alpha_u \in [0, 1]$. The yellow dot represents the starting point $\boldsymbol{\theta}^{(0)}$, whereas the blue dot represents the sought minimum*

`Adam` (with standard choice of $\beta_1 = 0.9$ and $\beta_2 = 0.999$) is definitely more robust than `SGD` and more effective than `Adadelta`, as shown in Figure 5.10, although the trend is opposite: `Adam` converges to $\hat{\boldsymbol{\theta}}_N$ for $\boldsymbol{\theta}^{(0-2)}$ (as shown in Figure 5.10(b)) and it gets somehow closer to it than `SGD`, for $\boldsymbol{\theta}^{(0-1)}$, as depicted in Figure 5.3.7.5, compared to Figure 5.8(a). Compared to standard `SGD`, `Adam` leverages the first- and second-order momenta and adapts their contribution to avoid divergence of the gradient descent algorithm. As a matter of fact, `SGD` needs (first-order) momentum too, to converge to the minimum in a reasonable amount of epochs (10,000), as depicted in Figure 5.8(a). This is evident if one compares the latter to the trajectory that `SGD` with zero momentum follows, as depicted in Figure 5.11(a) (with τ=0.5 to avoid divergence). The lack of momentum leaves the `SGD` algorithm iterating around local minima. However, a too-high momentum γ can lead the `SGD` algorithm to miss $\hat{\boldsymbol{\theta}}_N$ because of its inertia, as shown in Figure 5.11(b) for γ=0.9. For practical purposes, this aspect represents the major disadvantage of using SGD, compared to Adam adaptivity and faster convergence rate.

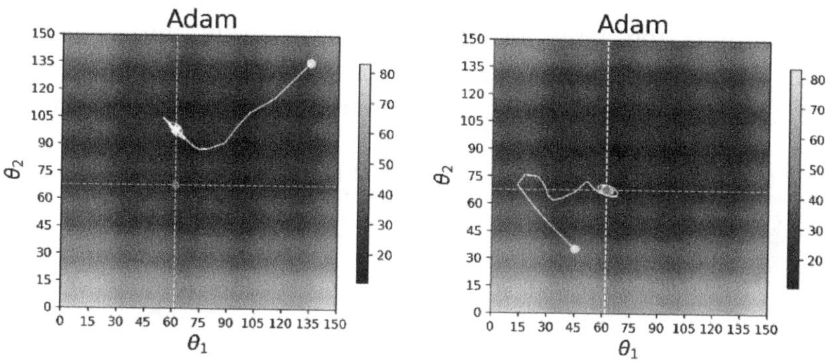

(a) (b)

Figure 5.10. *Adam minimization trajectory for $\theta^{(0-1)}$ (a) and $\theta^{(0-2)}$ (b). The yellow dot represents the starting point $\theta^{(0)}$, whereas the blue dot represents the sought minimum*

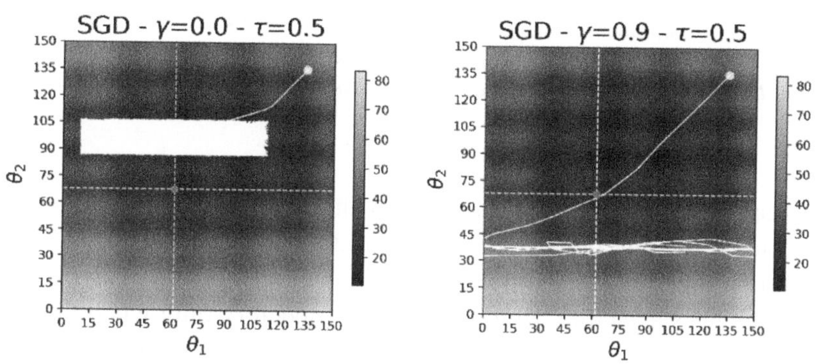

(a) (b)

Figure 5.11. *SGD minimization trajectory for γ = 0 and τ = 0.5 (a) and for γ = 0.9 and τ = 0.5. The yellow dot represents the starting point $\theta^{(0-1)}$, whereas the blue dot represents the sought minimum*

Nesterov momentum enhances the momentum effect (no damping is allowed in this algorithm), possibly leading SGD to reach the minimum $\hat{\theta}_N$ faster (in less epochs). However, as shown in Figure

5.12 for different initializations $\boldsymbol{\theta}^{(0-1)}$ (in Figure 5.12(a)) and $\boldsymbol{\theta}^{(0-2)}$ (in Figure 5.12(b)), Nesterov-SGD algorithm tends to range around the minimum $\hat{\boldsymbol{\theta}}_N$, but missing it because of the inertia effect (each final step represents a too large increment that iteratively misses the minimum). In this case, a proper schedule for the learning rate must be adopted. `RMSprop` is also conceived to counteract aggressive weight

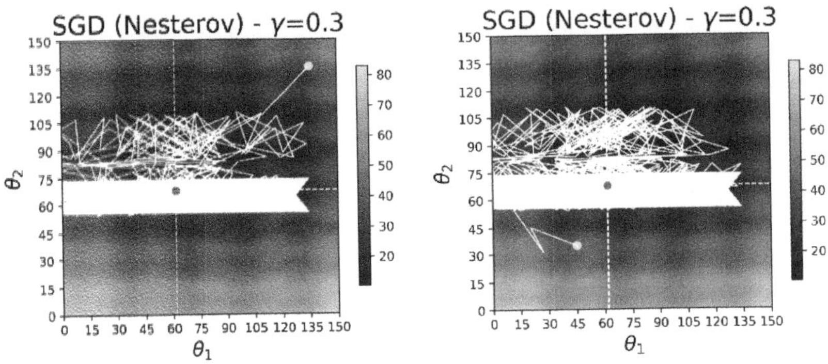

(a) (b)

Figure 5.12. *`SGD` minimization trajectory for $\boldsymbol{\theta}^{(0-1)}$ (a) and $\boldsymbol{\theta}^{(0-2)}$ (b). `SGD` is featured by Nesterov momentum γ = 0.3 and no dampening. The yellow dot represents the starting point $\boldsymbol{\theta}^{(0)}$, whereas the blue dot represents the sought minimum*

updates based on the estimated variance of the gradients computed in past iterations. Despite being more stable and effective than `SGD` with (Nesterov) momentum, for both weight initializations, in this example `RMSprop` achieves a "stable" gradient descent trajectory only when the momentum γ is compensated by the coefficient α_r and both are equal to 0.5 (see Figure 5.13(b)). However, (Mnih et al. 2016) showed that RMSprop is quite useful for very non-stationary problems, as confirmed by (Arjovsky et al. 2017), provided that no momentum is added (γ = 0), α_r = 0.9 and the learning rate is very small. Reducing the learning rate to 12, the minimum is reached with a clear improvement in the stability of the gradient descent trajectories, for both initializations, as shown in Figure 5.14. However, the instability close to the minimum $\hat{\boldsymbol{\theta}}_N$ persists. From those examples, it looks that calibrating `SGD`, `Adadelta`, `RMSprop` to find a reasonable trade-off

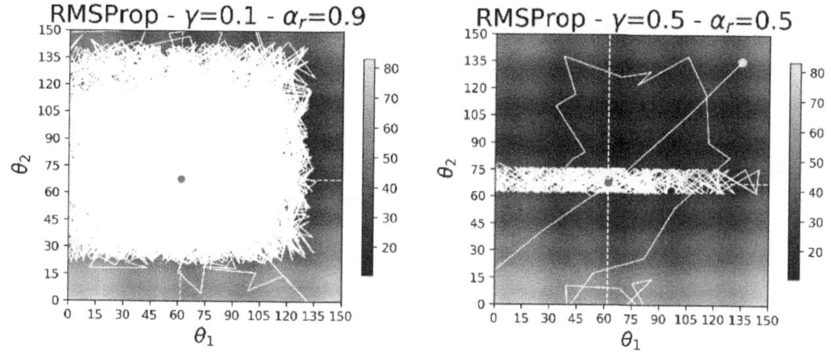

(a) (b)

Figure 5.13. *`RMSprop` minimization trajectory for γ = 0.1 and α_r = 0.9 (a) and for $\gamma = \alpha_r$ = 0.5 (b). The yellow dot represents the starting point $\theta^{(0-1)}$, whereas the blue dot represents the sought minimum*

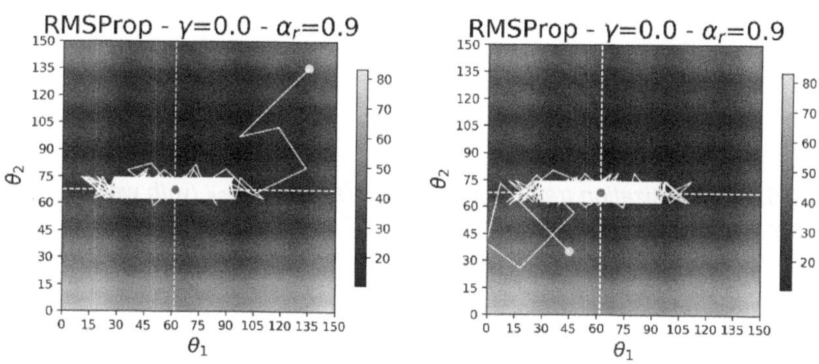

(a) (b)

Figure 5.14. *`RMSprop` minimization trajectory for $\theta^{(0-1)}$ (a) and $\theta^{(0-2)}$ (b). `RMSprop` is featured by γ = 0.0 and α_r = 0.9 for both cases. The yellow dot represents the starting point $\theta^{(0)}$, whereas the blue dot represents the sought minimum*

between convergence rate and stability is hard: either a small learning rate is assumed, with zero momentum and a very large number of epochs demanded to reach the minimum, either the dampened momentum is leverage, with high a risk of divergence. Standard `Adam`, on the contrary, seems more robust and flexible to complex optimization tasks. As far as weight decay is concerned, `Adam` (no weight decay) and `AdamW` (with weight decay) are compared in Figure 5.15. The use of `AdamW` does not provide an improvement, compared to standard `Adam`, despite the extremely low weight decay penalty equal to 0.00001. `Adam` still represents, for this example, the most robust and efficient optimizer, provided that no tuning is necessary, compared to other optimization techniques.

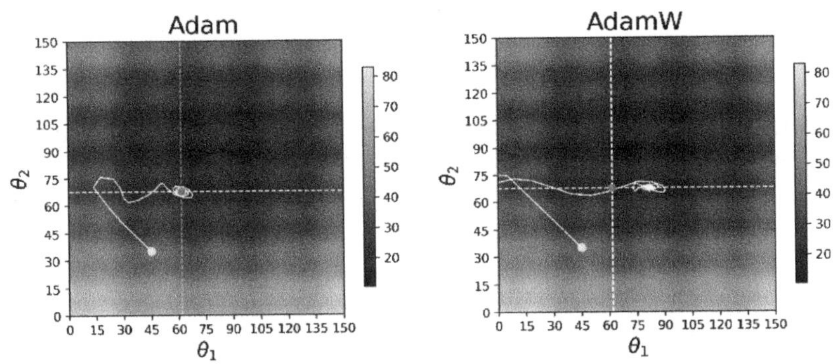

(a) (b)

Figure 5.15. *Minimization trajectories for* `Adam` *(a) and* `AdamW` *(with weight decay penalty of 0.00001) (b). The yellow dot represents the starting point* $\boldsymbol{\theta}^{(0-2)}$*, whereas the blue dot represents the sought minimum*

5.4. References

Arjovsky, M., Chintala, S., Bottou, L. (2017). Wasserstein generative adversarial networks. In *International Conference on Machine Learning*, PMLR, 214–223.

Bertsekas, D. (2015). *Convex Optimization Algorithms*. Athena Scientific, Nashua, NH.

Billingsley, P. (1995). *Measure and Probability*. John Wiley and Sons, New York.

Boyd, S.P. and Vandenberghe, L. (2023). *Convex Optimization*, 29th ed. Cambridge University Press, Cambridge.

Campagne, J.-E. (2018). L'apprentissage face à la malédiction de la grande dimension. Notes and comments on the lectures by S. Mallat of the Collège de France [Online]. Available at: https://github.com/jecampagne/cours_mallat_cdf.

Campagne, J.-E. (2019). L'apprentissage par réseaux de neurones profonds. Notes and comments on the lectures by S. Mallat of the Collège de France [Online]. Available at: https://github.com/jecampagne/cours_mallat_cdf.

Campagne, J.-E. (2020). Modèles multi-échelles et réseaux de neurones convolutifs. Notes and comments on the lectures by S. Mallat of the Collège de France [Online]. Available at: https://github.com/jecampagne/cours_mallat_cdf.

Campagne, J.-E. (2022). Information et complexité. Notes and comments on the lectures by S. Mallat of the Collège de France [Online]. Available at: https://github.com/jecampagne/cours_mallat_cdf.

Chouzenoux, E. and Pesquet, J.-C. (2019). Official handbook of the optimization. Lecture notes, "Ingénieur" Curriculum, CentraleSupélec [Online]. Available at: https://www-syscom.univ-mlv.fr/~chouzeno/ECP/index.htm.

Clouteau, D. (2022). Statistical learning for mechanics. Lecture notes, Cours de Modélisation des Incertitudes et Fiabilité des Ouvrages-Master 2 - Université Paris-Saclay, CentraleSupélec.

Cournède, P.-H. (2020). Official handbook of the statistics and learning. Lecture notes, "Ingénieur" Curriculum, CentraleSupélec.

Duchi, J., Hazan, E., Singer, Y. (2011). Adaptive subgradient methods for online learning and stochastic optimization. *J. Mach. Learn. Res.*, 12, 2121–2159.

Fisher, R.A. (1922). On the mathematical foundations of theoretical statistics. *Philosophical Transactions of the Royal Society of London. Series A*, 222, 309–368. doi: 10.1098/rsta.1922.0009.

Ioffe, S. and Szegedy, C. (2015). Batch normalization: Accelerating deep network training by reducing internal covariate shift. *arXiv* [Online]. Available at: http://arxiv.org/abs/1502.03167.

Jaynes, E.T. (1957). Information theory and statistical mechanics. *Phys. Rev.*, 106(4), 620–630. doi: 10.1103/PhysRev.106.620.

Kingma, D.P. and Ba, J. (2015). Adam: A method for stochastic optimization. *arXiv* [Online]. Available at: http://arxiv.org/abs/1412.6980.

Krizhevsky, A., Sutskever, I., Hinton, G.E. (2017). Imagenet classification with deep convolutional neural networks. *Communications of the ACM*, 60(6), 84–90. doi: 10.1145/3065386.

LeCun, Y., Bottou, L., Orr, G.B,. Müller, K.R. (1998). Efficient backprop. In *Neural Networks: Tricks of the Trade*, Orr, G.B. and Müller, K.R. (eds). Springer, Berlin, Heidelberg, 9–50 [Online]. Available at: http://link.springer.com/10.1007/3-540-49430-8_2.

Loshchilov, I. and Hutter, F. (2019). Decoupled weight decay regularization. *arXiv* [Online]. Available at: http://arxiv.org/abs/1711.05101.

Mnih, V., Badia, A.P., Mirza, M., Graves, A., Lillicrap, T., Harley, T., Silver, D., Kavukcuoglu, K. (2016). Asynchronous methods for deep reinforcement learning. In *Proceedings of the 33rd International Conference on Machine Learning*, PMLR, 48, 1928–1937.

Mula, O. (2019). Optimisation et programmation dynamique. Master mention Mathématiques Appliquées, 1ère année Université Paris Dauphine [Online]. Available at: https://www.ceremade.dauphine.fr/~mula/docs/teaching/optimM1/docs/ notes_optim.pdf.

Murphy, K.P. (2022). *Probabilistic Machine Learning: An Introduction*. MIT Press, Cambridge.

Nesterov, Y. (1983). A method for unconstrained convex minimization problem with the rate of convergence $O(1/k^2)$. *Dokl. Akad. Nauk. SSSR*, 269(3), 543.

Nielsen, F. (2020). An elementary introduction to information geometry. *Entropy*, 22(10), 1100 [Online]. Available at: https://www.mdpi.com/1099-4300/22/10/1100.

Nielsen, F. (2021). On a variational definition for the Jensen-Shannon symmetrization of distances based on the information radius. *Entropy*, 23(4), 464 [Online]. Available at: https://www.mdpi.com/1099-4300/23/4/464.

Nielsen, F. (2023). A simple approximation method for the Fisher-Rao distance between multivariate normal distributions. *Entropy*, 25(4), 654 [Online]. Available at: https://www.mdpi.com/1099-4300/25/4/654.

Peyré, G. (2020). Course notes on optimization for machine learning. Lecture notes, École Normale Supérieure [Online]. Available at: https://mathematical-tours.github.io.

Polyak, B. (1964). Some methods of speeding up the convergence of iteration methods. *USSR Computational Mathematics and Mathematical Physics*, 4(5), 1–17 [Online]. Available at: https://linkinghub.elsevier.com/retrieve/pii/0041555364901375.

Shimodaira, H. (2000). Improving predictive inference under covariate shift by weighting the log-likelihood function. *Journal of Statistical Planning and Inference*, 90(2), 227–244 [Online]. Available at: https://linkinghub.elsevier.com/retrieve/pii/S0378375800001154.

Smith, L.N. (2017). Cyclical learning rates for training neural networks. In *2017 IEEE Winter Conference on Applications of Computer Vision (WACV)*. IEEE, Santa Rosa, CA [Online]. Available at: http://ieeexplore.ieee.org/document/7926641/.

Smola, A.J., Vishwanathan, S.V.N., Hofmann, T. (2005). Kernel methods for missing variables. In *International Workshop on Artificial Intelligence and Statistics*, PMLR, 325–332.

Sutskever, I., Martens, J., Dahl, G., Hinton, G. (2013). On the importance of initialization and momentum in deep learning. In *Proceedings of the 30th International Conference on Machine Learning*, PMLR, Atlanta, GA [Online]. Available at: https://proceedings.mlr.press/v28/ sutskever13.html.

Tieleman, T. and Hinton, G. (2012). Lecture 6.5-rmsprop, COURSERA: Neural networks for machine learning. Technical Report 6, University of Toronto.

Vialle, S. (2022). Séquence Thématique ST7-76 - Similation à haute performance (2021-2022). Lecture notes, "Ingénieur" Curriculum, CentraleSupélec.

Wiesler, S., Richard, A., Schluter, R., Ney, H. (2014). Mean-normalized stochastic gradient for large-scale deep learning. In *2014 IEEE International Conference on Acoustics, Speech and Signal Processing (ICASSP)*. IEEE, Florence [Online]. Available at: http://ieeexplore.ieee.org/document/6853582/.

Zeiler, M.D. (2012). Adadelta: An adaptive learning rate method. *arXiv* [Online]. Available at: http://arxiv.org/abs/1212.5701.

List of Authors

Félix DARVE
3SR Laboratory
CNRS
Grenoble INP – ENSE3
Université Grenoble Alpes
France

Filippo GATTI
Laboratoire de Mécanique
Paris-Saclay, UMR 9026
CentraleSupélec
CNRS
ENS Paris-Saclay
Université Paris-Saclay
Gif-sur-Yvette
France

Noel JAKSE
SIMaP
CNRS
Grenoble INP
Université Grenoble Alpes
France

Filippo MASI
Sydney Centre in Geomechanics and
Mining Materials
School of Civil Engineering
University of Sydney
Australia

Ioannis STEFANOU
GeM, UMR 6183
CNRS
Ecole Centrale Nantes
Nantes Université
France

Index

I, K, L

M, N

P, R

S, T

U, W

Summary of Volume 2

Printed and bound by CPI Group (UK) Ltd, Croydon, CR0 4YY

13/09/2026

14961071-0001